DECODING THE UNIVERSE

ALSO BY
CHARLES SEIFE

Alpha & Omega
Zero

Charles Seife

DECODING THE UNIVERSE

HOW THE NEW SCIENCE OF INFORMATION IS
EXPLAINING EVERYTHING IN THE COSMOS,
FROM OUR BRAINS TO BLACK HOLES

VIKING

VIKING

Published by the Penguin Group

Penguin Group (USA) Inc., 375 Hudson Street, New York, New York 10014, U.S.A.

Penguin Group (Canada), 90 Eglinton Avenue East, Suite 700, Toronto, Ontario, Canada M4P 2Y3 (a division of Pearson Penguin Canada Inc.) • Penguin Books Ltd, 80 Strand, London WC2R 0RL, England • Penguin Ireland, 25 St. Stephen's Green, Dublin 2, Ireland (a division of Penguin Books Ltd) • Penguin Books Australia Ltd, 250 Camberwell Road, Camberwell, Victoria 3124, Australia (a division of Pearson Australia Group Pty Ltd) • Penguin Books India Pvt Ltd, 11 Community Centre, Panchsheel Park, New Delhi—110 017, India • Penguin Group (NZ), Cnr Airborne and Rosedale Roads, Albany, Auckland 1310, New Zealand (a division of Pearson New Zealand Ltd) • Penguin Books (South Africa) (Pty) Ltd, 24 Sturdee Avenue, Rosebank, Johannesburg 2196, South Africa

Penguin Books Ltd, Registered Offices: 80 Strand, London WC2R 0RL, England

First published in 2006 by Viking Penguin, a member of Penguin Group (USA) Inc.

10   9   8   7   6   5   4   3   2   1

Copyright © Charles Seife, 2006
All rights reserved

Drawings by Matt Zimet

ISBN 0-670-03441-X

Printed in the United States of America • Set in Adobe Minion • Designed by Amy Hill

CONTENTS

DECODING THE UNIVERSE

INTRODUCTION

Everything is made of one hidden stuff.

—Ralph Waldo Emerson

Civilization is doomed.

That's probably not the first thing you want to read when you pick up a book, but it's true. Humanity—and all life in the universe—is going to be wiped out. No matter how advanced our civilization becomes, no matter if we develop the technology to hop from star to star or live for six hundred years, there is only a finite time left before the last living creature in the visible universe will be snuffed out. The laws of information have sealed our fate, just as they have sealed the fate of the universe itself.

The word *information* conjures visions of computers and hard drives and Internet superhighways; after all, the introduction and popularization of computers came to be known as the information revolution. However, computer science is only a very small aspect of an overarching idea known as information theory. While this theory does, in fact, dictate how computers work, it does much, much more than that. It governs the behavior of objects on many different scales. It tells how atoms interact with each other and how black holes swallow stars. Its rules describe how the universe will die, and they illuminate the structure of the entire cosmos. Even if there were no such thing as a computer, information theory would still be the third great revolution of twentieth-century physics.

The laws of thermodynamics—the rules that govern the motion of atoms in a chunk of matter—are, underneath it all, laws about information. The theory of relativity, which describes how objects behave at extreme speeds and under the strong influence of gravity, is actually a theory of information. Quantum theory, which governs the realm of the very small, is a theory of information as well. The concept of information, which is far broader than the mere content of a hard drive, ties together all these theories into one incredibly potent idea.

Information theory is so powerful because information is physical. Information is not just an abstract concept, and it is not just facts or figures, dates or names. It is a concrete property of matter and energy that is quantifiable and measurable. It is every bit as real as the weight of a chunk of lead or the energy stored in an atomic warhead, and just like mass and energy, information is subject to a set of physical laws that dictate how it can behave—how information can be manipulated, transferred, duplicated, erased, or destroyed. And everything in the universe must obey the laws of information, because everything in the universe is shaped by the information it contains.

The idea of information was born from the ancient art of codemaking and codebreaking. The ciphers that hid state secrets were, in fact, methods of obscuring information and transporting it from place to place. When the art of code cracking was combined with the science of thermodynamics—the branch of physics that describes the behavior of engines, the exchange of heat, and the production of work—information theory was the result. This new theory of information was an idea as revolutionary as quantum theory and relativity; it instantly transformed the field of communications and paved the way to the computer age, but that was just the beginning. Within a decade, physicists and biologists began to understand that the ideas of information theory govern much more than the bits and bytes of computers and codes and communications: they describe the behavior of the subatomic world, all life on Earth, and even the universe as a whole.

Each creature on Earth is a creature of information; information sits at the center of our cells, and information rattles around in our brains. But it's not just living beings that manipulate and process information. Every particle in the universe, every electron, every atom, every particle not yet discovered, is packed with information— information that is often inaccessible to us, but information nonetheless, information that can be transferred, processed, and dissipated. Each star in the universe, each one of the countless galaxies in the heavens, is packed full of information, information that can escape and travel. That information is always flowing, moving from place to place, spreading throughout the cosmos.

Information appears, quite literally, to shape our universe. The motion of information may well determine the physical structure of the cosmos. And information seems to be at the heart of the deepest paradoxes in science—the mysteries of relativity and quantum mechanics, the origin and fate of life in the universe, the nature of the ultimate destructive power of the black hole, and the hidden order in a seemingly random cosmos.

The laws of information are beginning to reveal the answers to some of the most profound questions of science, but the answers are, in some ways, more disturbing and more bizarre than the paradoxes they solve. Information leads to a picture of a universe speeding toward its own demise, of living creatures as slaves to parasites within, and of an incredibly byzantine cosmos made up of an enormous collection of parallel universes.

The laws of information are giving physicists a way to understand the darkest mysteries that humanity has ever pondered. Yet those laws are painting a picture that is as grim as it is surreal.

CHAPTER 1

REDUNDANCY

Gentlemen don't read other gentlemen's mail!

—Henry L. Stimson

"AF is short of water." These five words sank the Japanese fleet.

In the spring of 1942, the U.S. military was reeling from an unbroken series of defeats. The Japanese navy was supreme in the Pacific, and it was pushing ever closer to American territories. Though the situation was dire, the war was not lost. And U.S. cryptanalysts were about to use a weapon as important as bombs and guns: information.

U.S. codebreakers had cracked JN-25, a cipher used by the Japanese navy. It was a tough code to break, but by May cryptanalysts had completely pried open the mathematical vault of the cipher and revealed the information hidden within.

According to the intercepted and decrypted messages, an American base, code-named *AF*, was shortly to be the object of a major naval assault. American analysts knew that AF was an island in the Pacific (quite possibly Midway Island), but they didn't know for certain which one it was. If the analysts guessed wrong, the navy would defend the incorrect island, and the enemy would be able to invade the true target unopposed. But if they could figure out which island AF really was and anticipate the destination of Japan's armada, the Americans could concentrate their fleet and maul the invading force. Everything—the war in the Pacific—hinged on one missing piece of information: Where was AF?

Commander Joseph Rochefort, head of the navy's cryptography center at Pearl Harbor, came up with a scheme to get that one last piece of information. He ordered the base at Midway to transmit a phony request for help. The transmission stated that the water distillery on Midway Island had been damaged and the base was nearly out of freshwater. The Japanese, who were eavesdropping on Midway's transmissions, heard the broadcast, too. This is precisely what Rochefort was counting on. Not long after the phony message, Navy Intelligence picked up the faint signals of a Japanese transmission on the airwaves: "AF is short of water." Rochefort had his last bit of information. AF was Midway.

The U.S. fleet gathered to defend the island. On June 4, 1942, Admiral Isoroku Yamamoto's invading force ran directly into Admiral Chester Nimitz's waiting fleet. During the battle, four Japanese aircraft carriers—the *Hiryu, Soryu, Akagi,* and *Kaga*—went to the bottom; in return, only one U.S. carrier was lost. The crippled Japanese fleet steamed home. Japan had lost the battle—and the war in the Pacific. The Japanese navy never again seriously threatened American territory, and the United States began the long, difficult drive to the Japanese homeland. A priceless piece of information, the target of Yamamoto's invasion, leaked through the protection of codes and ciphers and gave America its crucial victory.[1]

World War II was the first information war. As U.S. cryptographers extracted information from the Japanese JN-25 and Purple ciphers, an

---

1. Ironically, Yamamoto himself would die because of a bit of information intercepted by the Allies. In April 1943, a signals intelligence group in Australia discovered that Yamamoto was going to fly to New Guinea to visit the troops there. A squadron of P-38 fighters was waiting and shot down the admiral's plane over Bougainville in the South Pacific.

elite group of British and Polish codebreakers unraveled Germany's (supposedly) uncrackable Enigma cipher. And just as information allowed the United States to defeat Japan, the Enigma information gave the Allies a way to defeat the Nazi U-boats that were choking Great Britain.

Just as the struggle over information left its imprint on the face of the war, the war left its imprint on the face of information. During World War II, cryptography began to change from an art to a science. The codebreakers in the sweaty code rooms in Hawaii and on a quaint estate in England would be the heralds of a revolution known as information theory.

Codemaking and codebreaking were always closely related to what would become the theory of information. However, for millennia, cryptographers and cryptanalysts had no idea that they were making tentative forays into an entirely new field of science. After all, encryption is older than science. Over and over again, since antiquity, monarchs and generals have relied upon the information hidden by the fragile security of a cipher or a hidden message—awkward attempts to circumvent the dangers of information transfer.

Codemaking goes back to the dawn of Western civilization. In 480 BC, ancient Greece was nearly conquered by the much stronger Persian Empire, but a secret message, hidden under the wax of a writing tablet, warned of an impending invasion. Alarmed at the message, the Greeks immediately began preparing for war. The forewarned Greeks roundly defeated the Persians at the battle of Salamis, ending the Persian threat and ushering in the golden age of Greece. But for that hidden message, the fragile collection of Greek city-states would not have been able to resist the much more powerful Persian navy; Greece would likely have become a Persian conquest, and Western civilization would have turned out quite differently.

Sometimes, a failed attempt to transfer information changes history, too. Heads have literally rolled because a secret message or a cipher has been discovered and decrypted. In 1587, Mary, Queen of Scots, went to the executioner's block because of a bad code. Mary, in prison, was forming a conspiracy to murder Queen Elizabeth and seize the English throne. But since all objects that went in and out of the prison were inspected, Mary had to resort to cryptography to keep in touch with her supporters. She and her coconspirators devised a code and traded little enciphered messages hidden in the bungs of beer barrels. Unfortunately for Mary, Sir Francis Walsingham, England's spymaster, discovered the messages and had them deciphered. He even planted a fake message from Mary to the conspirators, inducing the traitors to reveal the names of all the men in their cabal. When Queen Mary stood trial for treason, the messages were the prime exhibit. A broken code—and two strokes of an axe—sealed her fate.

Codes and ciphers have many different forms, but they all have the same purpose: to transfer information from one person to another. At the same time, they must be secure; they must prevent an eavesdropper from getting that information if the message is intercepted.

Through most of history, codes weren't terribly secure. A smart codebreaker could unravel even the most sophisticated code with just a little bit of concentration; even so, monarchs and generals had to rely upon these rickety codes. Often, an intercepted and decrypted message meant death or defeat. Sending sensitive messages was always dangerous, but it was a necessary risk to take and a fundamental part of the business of diplomacy and war.

No matter how cryptographers fiddle with words or symbols or numbers or codebooks, no matter how cleverly they hide the messages in bungholes or pumpkins or within poems, there is an unavoidable risk of discovery as crucial information moves from place to place. Just as generals must move troops and arms and supplies from home to the front and back again, so too must they transfer information. And, in its way, information is every bit as palpable as the weight of a bullet, every bit as tangible as the heft of an artillery shell—and every bit as vulnerable as a freighter full of ammunition.

This fundamental property is the hardest thing to accept about information: information is as real and concrete as mass, energy, or temperature. You cannot see any of these properties directly, but you accept them as real. Information is just as real. It can be measured and manipulated just as the weight of an apple can be gauged with a scale or redistributed with a knife. This is why leaders, generals, and diplomats always took such risks with rickety ciphers. Information must travel from the sender to the receiver just as a hunk of gold bullion would have to travel from Fort Knox to the Mint. There's no magical way of transmitting the information instantly, just as there's no way to teleport the gold directly from vault to vault. Even the most advanced computers must find a way to transfer information from place to place—it can go over a telephone line or through a coaxial cable or even through the air via a wireless connection—but if you want to transfer information from computer to computer, it must travel physically somehow from one computer to another.

Because the information in an object is a concrete, measurable property like mass, this means that information can be misplaced or stolen in the same way mass can be. Just as someone who wishes to move gold from one place to another must brave the risks of highwaymen or thieves, a leader who wants to exchange information must brave the risks of interception and decryption. Information, like gold, must be moved around in order to have any value to humans.

Underneath all the cloak-and-dagger frillery, good codemakers and codebreakers are experts in manipulating information. A cryptographer designing a cipher is trying to ensure that information gets from a sender to a receiver without allowing anyone else to access that information; the information must not "leak" out of the encrypted message. Conversely, a codebreaker who intercepts an enemy's message is trying to extract information from a seemingly meaningless jumble of letters or symbols. This can only work if the cipher is imperfect—if information leaks out despite the codemaker's best efforts. But not even the best codemaker can make a message miraculously appear at the place it is needed; it must be transported. That's where it is most at risk of discovery.

This idea that something as seemingly abstract as information is actually measurable—and tangible—is one of the central tenets of information theory. This theory was born in the years right after World War II, when mathematicians laid out a set of rules that defined information and described its behavior. This theory has a mathematical certainty that is seldom seen in the sloppy, experimental world of science; its tenets are as inviolable as the laws of thermodynamics that prevent inventors from building a perpetual motion machine. Even though information had been around for centuries, it was only during World War II that cryptographers began to feel around the edges of information theory.

The science of cryptography holds the first clues to the nature of information. It will not yield the full story, but it will give an idea of how information is real and measurable and must be carried from place to place like a brick of gold. For one of the banes of a cryptographer—redundancy—is closely related to the concept of information, and understanding redundancy can help explain why information can be as palpable as an atom in a chunk of matter.

Whenever you receive a message, even something as simple as "the sky is blue," you must take the series of words and process them to understand the meaning of the message. You receive a series of marks on paper (or sounds in the air) and extract the meaning encoded in those marks. Your brain takes the raw set of lines and curves that spell out "the sky is blue" and manipulates those symbols until it understands that the message is a statement about the color of the heavens outside. This process, this extraction of meaning from a set of symbols, is an unconscious one. It's just something a human brain has been training to do from the very moment that parents make goo-goo noises at a baby in the cradle; the process of becoming fluent in a language is, in a sense, nothing more than learning how to pull meaning from symbols. However, this unconscious process—taking a stream of symbols and extracting meaning from it—is crucial to our ability to use language. And so is the concept of redundancy, because it is redundancy that makes a message easy to understand.

Redundancy is the extra clues in a sentence or a message that allow the meaning to be understood even when the message is somewhat garbled. As it turns out, every sentence in any language is highly redundant. A sentence of English—or of any other language—always has more information than you need to decipher it. This redundancy is easy to see. J-st tr- t- r--d th-s s-nt-nc-. The previous sentence was extremely garbled; all the vowels in the message were removed.[2] However, it was still easy to decipher it and extract its meaning. The meaning of a message can remain unchanged even though parts of it are removed. This is the essence of redundancy.

To humans, redundancy is a good thing, because it makes a message easier to comprehend, even when the message is partially scrambled by the environment. You can still understand a friend speaking in a crowded restaurant or talking on a staticky cell phone because of redundancy. Redundancy is a safety mechanism; it makes sure that a message gets through even if it gets slightly damaged in transit. All languages have these built-in safety nets composed of patterns and structures and sets of rules that make them redundant. You aren't usually aware of those rules, but your brain unconsciously uses them as you read, speak, listen, and write—anytime you are receiving a message from somebody in a natural language. Even though these rules aren't obvious, they are there nonetheless, and you can feel their influence if you play around with language a little bit.

Consider, for example, the nonsense word *fingry*. *Fingry* sounds as if it could be an English word. In fact, it sounds like an adjective. ("Gee, Bob, your boss looks like he's mighty fingry today.") But what if

2. This is also why shorthand schools could advertise their courses with signs that said, "If u cn rd th ad u cn gt btr jb & mo pa."

I create another nonsense word: *trzeci*? Unlike *fingry*, *trzeci* doesn't sound like a valid English word at all.[3] This is because of these implicit rules—in this case, rules specific to the English language. The letter *z* is fairly rare in English and never follows the letters *tr*. Furthermore, it's pretty uncommon to end a word with *i*, so *trzeci* doesn't feel like a real English word—it breaks the unwritten rules about the attributes of valid English words. *Fingry*, on the other hand, has the right pattern of letters (and sounds) to make it seem like a real English word, and the ending -*gry* tends to signal that the word is an adjective.

The human brain automatically learns these rules and uses them to do a validity check on all the messages that it receives. This is how we distinguish a message with meaning from a meaningless string of symbols or syllables.

All languages have rules within rules within rules. The *trzeci* versus *fingry* rules operate on the level of letters and sounds; they determine what letters and sounds are likely to follow others. But lots of other rules operate on different levels as well. Though they all function unconsciously, you can sense them when something is wrong with a message, because they automatically raise an alarm. For example, there are rules that determine what words are likely to follow other words and phrases; your brain, continuously monitoring the language rules, lets you know if words order are wrong used in the. There are also rules that check on the meaning of a message as you process it. Even a perfectly valid sentence will sound strange if it is not precisely what your brain expects. When this happens, an ill-chosen word can stick out like a sore earlobe.[4]

These rules are everywhere. They tell you the difference between a meaningless grunt and a meaningful consonant, between a nonsense word and a real one, or a silly sentence and one full of information.

---

3. Though it's a perfectly valid word under the rules of Polish. It means "third."
4. A cliché is nothing more than an overused—and highly redundant—turn of phrase. Just as you can replace the vowels in a sentence where they've been removed, you can often reinsert a missing word if it's just what the doctor . . .

Some rules are valid across many human languages; there are only a handful of sounds that are allowed to carry meaning in human speech. Some rules are more specific to individual languages; Polish words look and sound very different from English ones because the corresponding "valid word" rules are very different. But all languages have a vast set of these rules, and it is this set of rules that gives a language its structure—and its redundancy.

When your brain raises an alarm about a broken rule, a word that does not sound English or a sentence that has the wrong word, it is telling you that the stream of letters (or sounds) you are receiving doesn't meet its expectation for a valid message. Something is out of place; something is garbled. By using these rules and working backward, your brain can often correct the problem, such as when a word is missspeledd. Without missing a beat, your brain applied the proper rules of spelling and corrected the garbled stream of symbols. You extracted the meaning of the sentence despite an error. This is nothing more than redundancy in action.

These rules are also what allowed you to read the no-vowels sentence. The implicit rules of the English language instantly told you that "th-s" was probably *this* rather than *thms* or even *thes*. Thanks to the rules, you can still extract the meaning of a message even if I pare away bits of sentences . . . so long as I haven't removed too much. But there is a point beyond which a sentence can no longer be garbled or compressed without losing comprehensibility. Pare away too many letters and you begin to lose the meaning in the message. When you get rid of all the redundancy in a string of letters, what is left is a concrete, measurable, incompressible nucleus. That is information: the central, irreducible something that sits at the heart of every sentence.

This is a rough definition, and it's not a complete one, but it's accurate. Information and redundancy are complementary; when you remove the redundancy from a string of letters, or symbols for that matter, what's left is information. Computer scientists are well aware of this irreducible nub in every message. It is important when writing, say, a program for compressing computer files. Compression programs squash files—such as the ones that contain the text of this book—so that they take less room on a hard drive or similar storage devices. These programs are extremely good, but there is little mystery about how they do their job: they work by removing (almost) all of the redundancy from a file, leaving the nub of a file behind. A standard commercial compression program might take a text file and squash it by more than 60 percent. But what's left over is incompressible. Run that compression program again and the file won't squash down any further. (Try it yourself!) It can't be made any smaller unless you are willing to lose some of the meaning of the message, some of the information in the text file. If someone tries to sell you a program that can make such incompressible nubs even smaller, call the FBI to report a case of fraud.

Computer scientists aren't the only people concerned with redundancy. A key challenge of cryptography is to remove or mask the redundancy in a message while retaining that essential information at its heart. No matter how the cryptographers or computer scientists try to mask or shrink a message, though, there's still an incompressible chunk that must travel from the sender to the recipient, whether the message is transmitted by radio or by wax tablet or by lights in the steeple of the Old North Church. This realization would revolutionize the field of physics. But first, information and redundancy revolutionized the field of cryptography and changed the course of world history.

Modern cryptographers think of their craft in terms of information and redundancy. The goal of a cryptographer, after all, is to generate a stream of symbols that is meaningful to the intended recipient—in a sense, the cryptographer is creating an artificial language. Unlike ordinary human languages, which are intended to share information freely, the cryptographer's cipher is intended to be meaningless to an eavesdropper. The information in the original message is still there in the encrypted version; however, it is hidden to those who don't know how to decipher the message. A good cipher shields information from those not authorized to understand it. A bad cipher lets the information leak through. Often, when a cipher fails, it fails because of clumsy redundancy.

You already know this if you are an amateur codebreaker. On the comics pages of many newspapers, you will find a little puzzle known as a cryptogram. It's usually a famous quotation that has been encrypted in a very straightforward way: each letter is substituted for another letter of the alphabet, yielding a string of nonsense. For example, you might see something like FUDK DK V NTPVFDOTPM KDIAPT GSHDJX KGUTIT. DF KUSYPH JSF FVWT IYGU FDIT FS ZNTVW DF. With a little practice, you can quickly decipher this sort of puzzle and extract the information it contains.

There are several ways to decode a cryptogram, and they all exploit the unwritten rules of English. Even though the information is garbled, these rules allow you to figure out what the message is. One of the rules is that if you have a letter standing alone, it's either an *A* or an *I;* no other single letter forms a valid word. Therefore, in the above cryptogram, the symbol V must represent either the letter *A* or the letter *I.* Another rule is that *E* tends to be the most frequent letter in the English language, so in the above sentence, the most frequent symbol— T—probably represents the letter *E.* Some other letters, such as *S,* and combinations of letters, such as *TH,* are relatively common, so they are almost certain to appear in a given message, while others such as *X* or *KL* are quite rare and may well be missing from a typical cryptogram. Stare at the cryptogram and play around with it for a little while and you will soon be able to decipher the message. The rules of English allow you to extract the information from the message even though it's been hidden. In other words, these rules give the message redundancy and allow you to break the cipher.[5]

Redundancy, the collection of those patterns and rules, is the enemy

---

5. The above cryptogram translates to: This is a relatively simple coding scheme. It should not take much time to break it.

of a secure code; it helps information leak through, and codemakers go through a great deal of effort to try to hide the redundancy in a message. This is the only way that a codemaker can have hope that a new cipher *might* be secure. Understanding this relationship among redundancy, information, and security is a cornerstone of cryptography, but before information theory was born, nobody really had a deep understanding of what lay beneath that relationship. Nobody understood the nature of information or redundancy. Nobody had a formal method to define them or measure them or manipulate them. As a result, even the most sophisticated coding schemes of the early twentieth century tended to be insecure. Even those that were thought to be uncrackable.

In February 1918, the German inventor Arthur Scherbius filed a patent for an "unbreakable" cipher machine that would soon become infamous the world over: Enigma. Enigma was an ingenious way of encrypting a message. It was so complex that most contemporary cryptographers and mathematicians thought that it was hopeless to even attempt to break it.

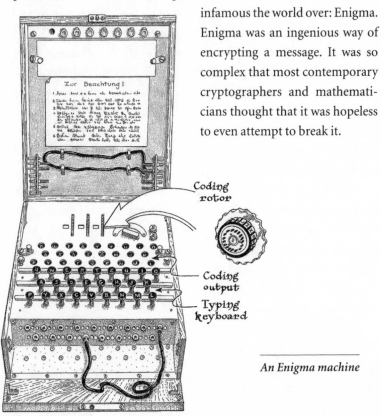

Coding
rotor

Coding
output

Typing
keyboard

An Enigma machine

Scherbius's machine looked somewhat like a sexed-up typewriter. However, a keystroke wouldn't make a mark on a paper; it would cause a light on the machine to turn on. If you pressed the letter key "A," for example, the light for the letter "F" might illuminate; the letter *A* was encrypted as an *F*. But if you pressed "A" again, it might show up as an "S" or an "O" or a "P"; each time you hit the letter "A," it would wind up encoded in a different way. This is because the heart of Scherbius's machine was a series of mechanical rotors. Every time you pressed a key, the rotors would turn, clicking forward one step. When the rotors changed position, the encryption changed, too. Every time you pressed a key it was encrypted in a different way. It was as if the Enigma machine changed ciphers with every keystroke.

Most models of Enigma used three rotors (though some had four), each of which would click forward twenty-six times before returning to its original orientation. These rotors could be wired in a number of different ways and placed in each of the three (or four) rotor slots. There were also wires and plugs that could be changed and some other features that could be altered, too. All told, a standard three-rotor Enigma machine could be configured in more than 300 million billion googol ways. If you were handed an encrypted Enigma message, you would have to figure out which one of those 3×10^{114} configurations the encrypter's machine was in when he started typing the message.

Brute force is out of the question; there's no way you could try each one of those 3×10^{114} configurations by hand. If every atom in the universe were an Enigma machine, and each one were trying a million billion combinations per second from the beginning of the universe until now, they still would only have been able to try 1 percent of all the possible configurations. No wonder Enigma had a reputation for being uncrackable. Luckily for Western civilization, it wasn't.

One of the best-kept secrets of the war was a small cadre of code-breakers at a Victorian estate: Bletchley Park in Buckinghamshire, England. Winston Churchill would later call the group the geese that laid golden eggs but never cackled. And Alan Turing was the most famous goose of them all.

Born in 1912 in London, Turing was to become one of the founders of the discipline of computer science—the field that deals with, in the abstract, objects that manipulate information. To mathematicians and computer scientists, Turing's most important contributions had to do with an idealized computer now known as a Turing machine, a mindless automaton that reads its instructions from a tape. This tape is divided into squares that either are blank or have a mark written on them. A Turing machine is extremely simple. It can only perform a handful of basic functions: read what's on the tape at a given position, advance the tape or rewind it, and write or erase a mark on the tape. In the 1930s, Turing and his colleague at Princeton University, Alonzo Church, proved that this simple robot is a *universal computer:* it can do any computation that a computer can conceivably do, even the most modern supercomputers. This means that you can, in theory, do the most complicated algorithms, the most intri-

A Turing machine

cate computerized tasks, if you are able to read, write, or erase a mark on a tape and move the tape around. This idea of a universal computer would be crucial to the development of computing and of information theory, but it is not for this that Turing is best known.

Turing is famous for cracking the Enigma cipher. Building upon work of Polish mathematicians, Turing and his colleagues at Bletchley Park exploited the redundancy in the coded Enigma messages to extract the information that the messages hid. Several flaws in the Enigma cipher inserted redundancy into the encoded message and helped weaken the cipher. Some of these flaws were due to its design. (For example, the Enigma machine would never leave a letter unchanged: an encrypted *E* could be any letter *except* an *E,* and this yielded a tiny bit of information about what the message was.) Some of these flaws were due to the Germans' method of communicating. (Bletchley Park codebreakers were able to exploit the predictability of encrypted weather reports to crack the code that hid those reports. And, like the predictability of language, this was a form of redundancy.) All together, the flaws allowed Turing and his colleagues to decrypt Enigma-enciphered messages with a series of primitive, specially built computing machines known as "bombes."[6] Turing and his fellow Bletchley Park geese eventually could crack an Enigma message in a matter of hours—a far cry from the billions and billions of years that a simple-minded analysis of the Enigma cipher's security would imply. Information leaked through the cipher; the Bletchley Park codebreakers were able to read that information even though it was hidden by the Enigma machine.

Just as cracking JN-25 changed the course of the Pacific war, cracking Enigma turned the tide of the Atlantic war. In the early stages of World War II, Germany's U-boat fleet nearly strangled the island fortress of Great Britain. Prime Minister Winston Churchill later wrote that "the only thing that ever really frightened me during the war was the U-boat peril." In the second half of 1940, the Nazi navy's "happy time," U-boats sent about half a million tons of shipping per month to the bottom of the Atlantic, nearly bringing Great Britain to her knees. The Enigma codebreakers changed that trend. Since U-boat

---

6. They got this name because they made an ominous ticking noise as they chugged away.

communications were encrypted with the naval version of Enigma, the Bletchley Park codebreakers helped British antisubmarine forces hunt down the U-boats that had caused their nation so much trouble, and helped win the war.[7]

The cryptanalysis of Enigma was the last great codebreaking effort before scientists learned how to define information, manipulate it, and analyze it. The Bletchley Park codebreakers, without really knowing it, were exploiting the irreducible, palpable nature of information. They were using redundancies, computer algorithms, and mathematical manipulations to burn through the cipher and extract the information that had to lie underneath it. In a sense, the cracking of Enigma was the shining star that heralded the birth of both computer science and information theory—and Turing's ideas would be an important part of both.

Sadly, Turing himself would not play a major role in the newborn science of information theory. In 1952, Turing, a homosexual, pleaded guilty to charges of "gross indecency" for his dalliance with a nineteen-year-old boy. To avoid imprisonment, he consented to undergo "treatment"—a set of hormone injections that were supposed to end his sexual proclivities. They didn't, and his "moral turpitude" was a stain that he never recovered from. Two years later, the tortured Turing apparently killed himself with cyanide.

Turing's tragedy came at the very moment when physicists and computer scientists would learn to deal with the entity of information, at a time when scientists would see that this hard-to-define concept of information holds the key to understanding the nature of the physical world. It was not the only suicide that cast its shadow on the science of information. In fact, tragedy lingered at the very roots of information theory, around the early work in physics that set the groundwork for the revolution to come.

---

7. Ironically, the U-boats had been helped as well as hurt by codebreaking. German codebreakers had cracked the Allied convoy code, allowing the German navy to send wolfpacks of U-boats to intercept Allied convoys.

CHAPTER 2

DEMONS

A hostile Demon are you, that I well perceive,
And fear your work is ever turning good to ill.

—Johann Wolfgang von Goethe,
Faust

On the afternoon of September 5, 1906, Ludwig Boltzmann found a small cord and wrapped one end around a crossbar in the wooden casement of a window. As his wife and daughter paddled happily about the bay of the resort town of Duino, in what was then Austria-Hungary, Boltzmann fashioned a crude noose with the other end of the cord and hanged himself. His daughter found the body.

Inscribed on Boltzmann's grave is a very simple equation: $S = k \log W$. This expression would revolutionize two seemingly unrelated fields of physics. The first, thermodynamics, deals with the laws that govern heat, energy, and work—and is the source of the most powerful law of physics. Boltzmann would not survive to see the second, information theory, come to life.

At first glance, thermodynamics and information theory might seem as if they have nothing in common. One deals with the extremely concrete ideas that any nineteenth-century engineer could appreciate. Heat. Energy. Work. These are the things that make factories run, steam engines chug, and foundries glow. Information, on the other hand, is seemingly evanescent and abstract; you can't put information in a vat and have it melt steel, or stick it in a mill and make it spin wool. Nevertheless, the roots of information theory lie in thermodynamics. And both disciplines are rife with demons.

In the late eighteenth century, Europe was a continent full of demons, and France was no exception. The French Revolution in 1789 deposed Louis XVI and eventually toppled the head from his shoulders, and in the despotic fervor in the years afterward, a great many citizens followed their monarch to the grave. Among them was the great French scientist Antoine-Laurent Lavoisier.

Lavoisier was partially responsible for the birth of the discipline now known as chemistry. His experiments showed that chemical reactions neither destroyed mass nor created it—when you burn something, for example, the mass of the products always equals the mass of the reactants—a principle now known as the conservation of mass. He also proved that the process of combustion was due to a substance in air: oxygen. In his *Elementary Treatise of Chemistry,* which was published the same year as the French Revolution, he set the groundwork for the new scientific field of chemistry, in part by listing a set of "elements," fundamental substances that could not be further divided. Oxygen was among them, as were hydrogen, nitrogen, mercury, and a number of others whose existence is now second nature to chemists. But one of Lavoisier's "elements" is unfamiliar to modern scientists: caloric.

Lavoisier, along with most of the scientists of his day, was convinced that caloric, an invisible fluid that could flow from object to object, was responsible for how hot or cold something was. A hot bar of iron, argued Lavoisier, was dripping with caloric, while a cool chunk of marble didn't have very much at all. If you put the iron in contact with the marble, the caloric fluid would, in theory, flow from the iron to the marble, cooling down the former and heating up the latter.

This idea is wrong, though Lavoisier himself didn't live to see the overthrow of caloric theory. An aristocrat, he was viewed with suspicion by the officials in the Reign of Terror, and they looked for a way to get rid of him. In 1794, he was arrested, charged, and convicted of watering down tobacco to cheat the people. On May 8, an appointment with the guillotine cut Lavoisier's promising career (and his neck) short.

His beautiful widow, Marie Anne, would remarry—and her new husband would eventually prove that Lavoisier's caloric was a fiction. Benjamin Thompson was born in Massachusetts in 1753, but had to flee the country because he was a spy for Britain, reporting on the goings-on of his revolutionary Colonial colleagues. He bounced around Europe, married (and divorced) Marie Anne Lavoisier, and wound up as a military engineer in Bavaria.

There was a great demand in turbulent Europe for artillery, and part of Thompson's job was to oversee the building of cannons. Workers would take a chunk of metal and bore out a hole in the barrel with a drill. Thompson noticed that a dull drill bit wouldn't bite into the metal and would continue grinding away without cutting—but it would generate heat. As the drill turned and turned, the cannon metal got hotter and hotter, and it would stay hot as long as the drill kept turning.

This didn't make any sense with caloric theory. If heat were caused by some sort of fluid flowing from the drill bit to the cannon barrel, then at some point the supply of fluid should be used up. Instead, the heating continues so long as the drill bit turns: it is as if the turning drill has an endless amount of caloric. How could a tiny drill bit contain an infinite amount of fluid?

Thompson's cannons showed that heat wasn't, in fact, caused by an invisible fluid. Instead, the drill bit was doing *work* by rubbing against the metal of the cannon, and that work was being converted into heat. (You do the same thing when you rub your hands together, and less obviously, when you shiver on a cold winter day. You are converting the work done by movement into warmth.) It would be a number of years before scientists fully realized that the phenomenon of heat and the work done by physical motion were intimately connected, but it was this realization that helped build a new scientific discipline known as thermodynamics.

Not all the revolutions in Europe were political. Just as kings were overthrown, so were ancient lifestyles and old ideas. The science of thermodynamics was born in a revolution that swept away the last vestiges of the feudal system: the Industrial Revolution. All over Europe, inventors and entrepreneurs were trying to automate labor-intensive tasks and create machines that were stronger and faster than humans and pack animals. The cotton gin, the power loom, the locomotive—all of these inventions didn't need wages and allowed industrialists to make ever-greater profits. But at the same time, these inventions needed power to make them work.

Before industrialization, human power, animals, and the flow of water were sufficient power sources for the machines of the day. But the machines of the Industrial Revolution required much more power than the machines of old, so the "engine" was born.[1] The most famous was patented in 1769 by the Scottish inventor James Watt: a sophisticated version of the steam engine.

In principle, the steam engine is very simple. First, you need a fire. This fire causes water to boil into steam, which takes up more room than the equivalent amount of water—it expands. The expansion of the steam does work: it moves a piston which then, in turn, can move a wheel or lift a rock or pump water. The steam then either flies away into the sky or moves into a cool chamber exposed to the air and then condenses, flowing back toward the fire to begin the cycle again.

Even more abstractly, the steam engine sits in between a high-temperature object (the fire) and a cold-temperature object (the air).

---

1. A century earlier, the word *engine* meant nothing more specific than "mechanical thingy." Industrialization gave us the specific meaning of an object that provides power.

It allows heat to flow from the high-temperature reservoir to the cold-temperature one through the motion of the steam. At the end of the cycle, the hot object is a little cooler (you have to keep stoking the fire to keep it going), and the cool object is a little warmer (the steam has heated the surrounding air a bit). But in allowing that heat to flow, the engine extracts some of the energy and performs useful work.[2] And so long as there is a temperature difference between the hot reservoir and the cold reservoir, an ideal engine like this—a heat engine—will continue to putter away.

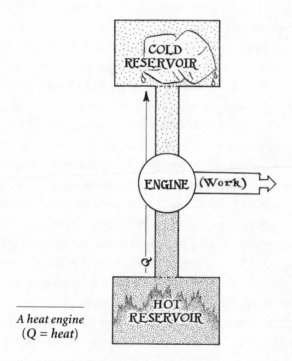

A heat engine
(Q = heat)

---

2. Lots of engines work in this way. The modern four-cycle gasoline engine, for example, is really a heat engine like this. The hot reservoir is the gasoline-air mixture right after it's ignited. The expansion of that mixture drives a piston and releases the hot gases into the cold reservoir (the air). From the physicist's point of view, it's little different from a steam engine.

Benjamin Thompson, the English physicist James Joule, and other scientists would show that there is a relationship between work and heat—that both work and heat are ways of transferring *energy*. There is energy stored in a lump of coal or a drop of gasoline. By burning them, you can release that energy and transfer it into the engine itself. The engine then uses some of that energy in doing useful work—lifting a concrete block a few meters, for example. But some of that energy is released into the environment. And unless you keep adding energy to the hot reservoir to keep it warm (or continuously remove energy from the cool reservoir to keep it cool—more on this shortly), the two reservoirs will soon reach the same temperature and the engine will sputter to a halt.

Obviously, engineers would like to use as much of that energy as possible in doing useful work, and waste as little of that energy as they can by minimizing the heat released into the environment. In other words, they want to make their engines as *efficient* as possible. This became a major effort; in the early 1800s, one of the big problems was figuring out how to make steam engines ever more efficient. It was a child of the French Revolution who discovered the ultimate limit to the power of an engine.

Sadi Carnot was born in Paris in 1796, two years after Lavoisier lost his head to the guillotine. His father, Lazare, was a general and a member of the pre-Napoleonic French government. The young Carnot became, like Benjamin Thompson, a military engineer, but his interests soon turned to the problem of steam engines. And he was more scientifically minded than Thompson: he wanted to find out the general principles that limited the engines of the engineers.

In the 1820s, scientists still had little understanding about the interconnections between heat, work, and energy in an engine, so Carnot began calculating, setting up careful analyses to figure out how the ideas interrelated. For example, in 1822 he tried to determine how much work can be done by a given amount of steam. But Carnot is most famous for figuring out how much work a steam engine *can't* do.

Carnot's brilliant idea was to examine an engine that is, in theory, totally reversible. Each of the steps in this (imaginary) engine's cycle can, the instant after completion, be reversed without any loss. For example, a quick and violent compression of a cylinder full of gas is reversible; if allowed to do so, the gas could expand to its original volume, pressure, and temperature, completely reversing the compression. It turns out that the efficiency of a reversible, Carnot, engine depends only on the temperatures of the heat reservoirs. Nothing else matters. For example, a Carnot engine that uses just-evaporated steam at 100 degrees Celsius and ejects the steam into the air on a freezing day at 0 degrees Celsius can only be about 27 percent efficient. Only about 27 percent of the energy stored in the steam can be turned into useful work; the rest flows as heat into the air.

That doesn't seem like a very efficient process. Three-quarters of the energy is wasted by a Carnot engine operating between 0 and 100 degrees Celsius. But it turns out that this is the most efficient heat engine you can get. Here's where the reversibility comes in.

A heat engine straddles a hot reservoir and a cold reservoir. By cycling through several steps, the engine allows heat to flow from the hot reservoir to the cold reservoir and, in the process, extracts useful work, say, by turning a crank. In a Carnot engine, though, every step is reversible. In fact, you can run the entire cycle backward. You can take a Carnot engine and put work into it. Turn the crank. This makes the cycle run in reverse. The engine pumps heat from the cold reservoir to the hot reservoir: the hot side gets hotter and the cool side gets cooler. A heat engine, in reverse, is a heat pump: put work in and you cool down a cold reservoir and heat up a warm reservoir.

Refrigerators and air conditioners are heat pumps like this. In refrigerators, the cold reservoir is inside the fridge, and after you add work with an electric motor, the pump takes heat from inside the fridge and releases it into the hot reservoir, the room-temperature air in your kitchen. With air conditioners, the cold reservoir is the room you are cooling; the hot reservoir is the summer's day outdoors, which is why you always must have a component of your air-conditioning system sticking out of the room you're cooling.

Now, imagine a Carnot heat engine and a Carnot heat pump straddling the same reservoirs. The Carnot engine allows a certain amount of heat, Q, to flow from the hot reservoir to the cold reservoir. In the process, it produces a certain amount of useful work. The Carnot heat pump consumes that work and, in so doing, pumps Q heat from the cold reservoir to the hot reservoir. Hook the heat engine and the pump together and they exactly cancel out each other. Looking at the engine-pump system as a whole, no net heat flows from reservoir to reservoir, and no net work is performed.

In 1824, Carnot realized that something very odd happens if you change the picture slightly. Imagine that you've got a superengine; it's more efficient than the Carnot engine operating under the same conditions. While allowing the same amount of heat, Q, to flow from the hot reservoir to the cold reservoir, it performs a little more work than the Carnot engine does. Pull the Carnot engine out of the engine-pump system and replace it with the superengine. Since the super-engine produces more work than the Carnot engine did—and what the Carnot heat pump consumes—you can divert a little bit of work away from the engine and still keep the heat pump running. The heat pump consumes the same work as before and pumps Q heat from the cold reservoir to the hot reservoir. All told, no net heat flows from the cold reservoir to the hot reservoir, but since the superengine produces a little more work than the Carnot engine did, some usable work is left over that isn't needed to run the Carnot heat pump. That is, you've created a heat engine that does work for free. It doesn't allow any net heat to flow from the hot reservoir to the cold reservoir, yet it still can lift rocks or move a locomotive. You've created a perpetual motion machine, since the engine does work without consuming anything (you don't need fuel to keep the hot reservoir hot) or changing the environment (the cool reservoir doesn't warm up, and the hot reservoir doesn't cool down).

But nothing comes for free. It's the law.

————

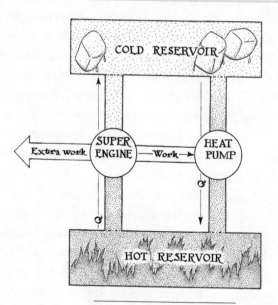

A perpetual motion machine

When Carnot formulated his argument about the efficiency of steam engines, it was the beginning of the science of thermodynamics. In the 1820s, scientists knew very little about heat, work, energy, and temperature; they were beginning to develop a sense about the interrelationship of all these ideas, but at the time they were ignorant of the most basic facts that physicists take for granted today. In Carnot's time, for example, nobody knew one of the most fundamental laws of our universe: energy cannot be created or destroyed. Energy is conserved. The amount of energy in the universe is a constant.

The first clue to this came not from steam engines, but from electric ones. In 1821, the British scientist Michael Faraday invented an electric motor. In the modern incarnation of such a motor, an electric current passes through a coiled wire that is surrounded by magnets. The magnetic field exerts a force on the current-carrying wire and causes it to spin—and you can use that spinning motion to drive a crank or do other useful work.

James Prescott Joule, son of a Manchester brewer, was experimenting with electric motors when he realized that current passing through a motor heats up the motor itself. But a motor that's doing useful work will generate less heat than a motor that's jammed and not spinning. Do more work, make less heat. Do less work, make more heat. Like Benjamin Thompson, Joule had found a connection between physical work—lifting rocks, turning drill bits—and the generation of heat. But unlike Thompson, Joule was a careful experimenter and set to work measuring precisely how much heat and how much work were generated under different conditions.

Joule did a large number of experiments with different systems, not only electric motors but also physical systems like waterwheels, and figured out how work is converted to heat is converted to electricity, and back again. For example, he dropped a weight and used the physical motion of the weight to turn a generator and create an electric current in a wire, showing the relationship between physical work and electrical energy. In his most famous experiment, he used the motion of a paddle wheel to warm up a container full of water, demonstrating, once and for all, that work can be converted into heat. Because they are interconvertible, work, heat, and electrical energy—in fact, all forms of energy—can be measured in the same units.

Just as the fundamental unit of time is the second and the fundamental unit of distance is the meter, the fundamental unit of energy is the joule. One joule will allow you to lift a one-kilogram rock by about one-tenth of a meter; it will heat a gram of water by about a quarter of a degree Celsius; it will light a one-hundred-watt electric lightbulb for one-hundredth of a second.

By his basement experiments, James Joule showed that work and heat were means of transferring energy from one body to another. If you lift a one-kilogram weight by one-tenth of a meter, the weight has one joule more energy than when you started; similarly, if you heat up a gram of water by a quarter of a degree, the water has one more joule of energy than when you began. He also showed that if you are really clever, you can convert from one form of energy to another; dropping a one-kilogram weight by one-tenth of a meter can, in theory, heat up a gram of water by a quarter of a degree. (In reality, you can never convert the full amount, as will become abundantly clear shortly.) But in all of these experiments, Joule realized that you never get more energy out of a system than you put in. A kilogram weight dropping a tenth of a meter will never, ever heat up a gram of water by *more* than a quarter of a degree. The energy doesn't appear out of nowhere. In the experiments, Joule was converting energy from one form to another, but he never was able to *create* energy.

Joule—and a number of other contemporary scientists—had discovered what is now known as the *first law of thermodynamics*.[3] Energy cannot be created. In fact, it cannot be destroyed, either. It can change forms; it can be transferred in the form of work or heat; it can dissipate; it can speed out of the room that you are experimenting in. But energy never simply pops into existence or is annihilated into nothingness.

This is an extremely powerful law. It tells you that the amount of energy in the universe is a fixed constant, that all the energy we will ever be able to use is already here, stored somewhere in another form. Whenever we use energy—to heat something or to do physical work—we are simply converting preexisting energy (the chemical energy stored in coal)[4] into a different form that is more useful to us. A steam

---

3. When physicists of the seventeenth and eighteenth centuries found a fundamental rule that the universe seemed to obey, they dubbed it a *law*. Many of these laws are profound and important, such as the laws of motion, the law of universal gravitation, and the laws of thermodynamics. Some laws are less deep—such as Hooke's law (which talks about how springs behave) or Snell's law (which describes how light bends when it moves from one medium to another). Modern physicists tend not to use the word *law*, as it implies an infallibility that isn't truly there when you examine the laws closely. That's why quantum mechanics and general relativity tend to be referred to as *theories* rather than laws, though the two terms can be used (more or less) interchangeably. (Theories also tend to refer to a framework, while a law is usually a single equation.)
4. So where does the energy in coal come from? Coal is highly compressed organic material, such as wood; the chemical energy is stored in carbon-based molecules. Wood is full of stored energy because the tree it came from took sunlight—a form of energy—

engine, for example, cannot create energy; it is extracting energy from its fuel. It's one of the most fundamental rules in physics: energy can be neither created nor destroyed. But there was an even more powerful law to come.

In the 1860s, the German physicist Rudolf Clausius noticed a subtle pattern to what energy transformations do to their environment. A heat engine relies upon a temperature difference—a hot reservoir and a cold reservoir—to work; it allows heat to flow from the hot side to the cold side and extracts work in the process. When the engine is done running, the hot side has cooled down and the cold side has heated up; the two reservoirs are closer in temperature than they were when the engine started running. The two sides started off very different, and the engine brought them closer to equilibrium with each other. In a sense, the equilibrium of the universe as a whole increases when you run an engine.

Can you get the two reservoirs *further away* from equilibrium instead of closer? Sure. All you need is a heat pump straddling the two sides. Add energy in the form of work, and the hot side gets hotter and the cold side gets colder; the two are brought ever further out of equilibrium. But Clausius realized that there was a hitch. How can you do the work to run the heat pump? Perhaps you run another engine—but that engine increases the equilibrium of the universe as it runs, canceling (in fact, *more* than canceling) the decrease in equilibrium caused by the heat pump. The equilibrium of the universe increases, despite your best efforts.

What if you don't use an engine? What if you turn a crank by hand?

---

and used it to convert water and carbon dioxide into energy-storing carbon-based molecules. So where did the sunlight come from? The sun is taking hydrogen atoms and fusing them together. The fusion of the two atoms releases energy that is stored in them (in the form of mass, as described by Einstein's theory of relativity). So where did the mass of the atoms come from? It came with our universe—from the big bang. So where did the big bang come from? Good question . . . and nobody's really sure, though there are some possible explanations. But all the energy (including Einsteinian mass-energy) currently in our universe was created with the big bang, and the amount hasn't changed since the universe's birth.

Well, in actuality your muscles are acting as an engine, too. They are exploiting the chemical energy stored in molecules in your bloodstream, breaking them apart, and releasing the energy into the environment in the form of work. This increases the "equilibriumness" of the universe just as severely as a heat engine does.

In fact, there's no way to get around the ever-growing equilibriumness of the universe. Whenever someone uses an engine or does thermodynamic work, the process automatically brings the universe closer to equilibrium. You can't counteract the increase in equilibriumness with a heat pump or another device, because the work needed to run the device would have to come from an engine or a muscle or some other source that cancels out the heat pump's efforts.[5]

This is the *second law of thermodynamics*. It is impossible to reduce the equilibriumness of the universe; in fact, every time you do work, you drive the universe closer to equilibrium. Where the first law says you can't win—you can't create energy out of nothing—the second law says you can't break even. Anytime you do useful work, you are irreversibly increasing the equilibriumness of the universe. The second law is also why there is no such thing as a superengine that works better than a Carnot engine. A superengine hooked up to a Carnot heat pump is doing useful work without changing its environment; in fact, you can isolate that engine-pump system in a box, yet it will still be able to keep doing useful work indefinitely. The equilibriumness of the universe wouldn't change at all, even though your machine can do useful work. But the second law of thermodynamics says that an engine or other device must feed on the nonequilibriumness of the universe—and you can't create work out of nothing, thanks to the first law of thermodynamics. Therefore, the superengine cannot exist. It would lead to a device that does work indefinitely without reducing the equilibrium of the universe. It would lead to a perpetual motion machine.

---

5. Those of you with a physics background might recognize that equilibriumness is really a way of saying *entropy*. More on this later in the chapter.

Inventors and hucksters have been trying to build perpetual motion machines for centuries, and even today there are lots of people who will try to sell you one. (Since "perpetual motion machine" is a sure way to scare off investors, the current term of the art is *above-unity device*.) Some of the designs are based upon magnetic fields; others are based upon various "quantum" technologies. The U.S. Patent Office has been so deluged with applications for perpetual motion machines that the office has a special rule for such devices: an inventor must submit a working model with his application. (Nevertheless, some slip through the cracks and actually attain a patent.) But the second law of thermodynamics—now considered to be the most unassailable law in physics—absolutely forbids the creation of a perpetual motion machine. Save your money for better investments, like ownership of the Brooklyn Bridge.

The second law of thermodynamics was a great victory for scientists of the mid-nineteenth century, but it marked a change in the tone of physics. Since Newton's time, physicists had been discovering laws about the universe that made humans more powerful. They learned how to predict the motions of the planets and of physical bodies, they were learning how matter behaved—and each discovery increased the number of things that clever scientists and engineers could do. The first and especially the second law of thermodynamics told them what they *couldn't* do. You can't create energy out of nothing. You can't do work without disturbing the universe. You can't build a perpetual motion machine. These were the first real, indisputable restrictions on human endeavor handed down by Nature. Despite its restrictive character, the second law is crucial for modern physics. Indeed, the physicist Arthur Eddington said that "The law that entropy always increases— the second law of thermodynamics—holds, I think, the supreme position among the laws of Nature." Physics was beginning to learn the limits of its own power, and these limits would become an important theme in the twentieth century.

In the 1860s, though, this change in tone only caused a vague unease as scientists hammered out the rules of thermodynamics, the physical principles that govern the interrelationship of energy, work, heat, temperature, and the nature of reversibility and irreversibility. But it would take the depressive Ludwig Boltzmann to solve the full mystery of the new branch of physics. With a new set of mathematical tools, Boltzmann figured out the reasons for the most fundamental laws known to physics. Boltzmann's work would change the way scientists looked at matter, temperature, and energy—and it would lay the foundations for the way they analyze information. He, too, would struggle with demons.

Ludwig Boltzmann was born in 1844 in Vienna, the son of a bureaucrat. Though socially clueless, the young man was an excellent student and in his early twenties began to tackle important problems at the forefront of physics. At the time, the forefront was atomic theory.

In the seventeenth century, scientists had figured out some general properties of gases. If you have a tube full of gas and use a piston to reduce the volume of the chamber to half its original size, the pressure of the gas inside doubles. That's a law discovered by the English chemist Robert Boyle. If, instead of squashing the gas, you double its temperature, it will push out the piston so that the volume of the chamber doubles. Heat a gas and it expands; cool it and it contracts: that is the essence of Charles's law, named after the French chemist Jacques Charles.

Thanks to generations of clever experiments, scientists had a pretty good sense about the interrelationship of the pressure, temperature, and volume of a gas in a container. But empirical knowledge doesn't always mean deep understanding. It was only in the mid-1800s that physicists started to understand *why* gases behaved in this way.

Modern physicists know that a gas, like helium, is made of tiny particles—atoms. These atoms are constantly in motion, flying about the container at different speeds. When an atom collides with the container, it ricochets away like a racquetball off a wall. But the collision gives the container wall a tiny knock. One collision has little effect on the walls of the container, but zillions upon zillions of these tiny ricochets collectively take their toll. They exert a strong pushing force on the container walls, forcing them outward. This is the source of a gas's pressure.

If you squash the container, then the same number of atoms are in a smaller space, and because the container is more crowded, more of these racquetball atoms slam against the walls per second. The number of ricochets goes up, the collective force that the ricochets exert increases, and the pressure rises. This is what causes Boyle's law: because decreases in volume increase the frequency of collisions and vice versa, volume and pressure are inversely proportional. Similarly, physicists now know that a gas's temperature is a measure of how much energy the atoms have. This, in turn, is related to how fast the atoms are skittering around. The hotter the gas is, the more energy it has and the faster the atoms move, on average. (This is the true nature of temperature. It is a measure of how energetically, and by extension how fast, an atom is moving. A hot atom of helium is moving faster than a cold atom of helium; conversely, a fast-moving atom is hotter than a slow-moving atom of the same species.) The more energetic an object is and the faster it moves, whether it be a racquetball or an atom or an SUV barreling down the highway, the more kick it imparts to the object it slams into. So, increase a gas's temperature and the atoms move faster and ricochet off the walls harder, giving the walls of the container an even firmer push—the pressure of the gas increases. If the walls of the container are allowed to move, the container will expand to equalize the pressure. This is what causes Charles's law.

Atomic theory ties pressure, temperature, volume, and energy—all the concerns of thermodynamics and the steam engines of the Industrial Revolution—into a nice, neat package. However, what is obvious to modern scientists was difficult to accept for nineteenth-century physicists. After all, nobody had the means to detect an individual atom; as late as the early twentieth century, some eminent scientists refused to believe that atoms existed at all. But in the mid-nineteenth century, physicists were beginning to realize that atomic theory—the idea that matter was composed of constantly moving, billiard-ball-like particles—did an excellent job of explaining the properties of gases and other types of matter. In 1859, Rudolf Clausius published a paper that set the groundwork for what became known as the *kinetic theory of gases*—but he ran into trouble. He couldn't get the numbers to work out quite right.

The trouble was with temperature. Clausius knew that temperature was a measure of the energy of the atoms in the gas: the hotter the gas was, the more energy the atoms had and the faster they were moving. Indeed, if you know how hot a gas is and how heavy the atoms are, then you can easily work out the speed of an average atom. Clausius did this and then worked out what would happen if you had a container full of tiny atomic billiard balls all moving at this particular speed. While the results were encouraging, Clausius's analysis wasn't quite correct; the relationship of pressure, temperature, volume, and energy wasn't quite what was observed in nature.

In 1866, the Scottish physicist James Clerk Maxwell figured out the flaw in Clausius's argument. While Clausius assumed that all the atoms in the gas were moving at the same speed, Maxwell realized that when the billiard balls collided with the walls and with each other, they exchanged energy. Some would wind up moving faster than average and others would wind up moving slower than average. Maxwell realized that by assuming that the speeds of the molecules had a particular *distribution,* he could fix the inaccuracies in Clausius's theory.

A distribution is an expression that appears often in a branch of mathematics, probability theory, that deals with uncertainty. It is a measure of how common something is. Imagine someone asks you how tall the average American adult male is. That's not too tough a question. You can say that the average height is around 5 feet 9 inches tall. But what happens if the person asks you to describe how tall American males are, in general? You can't just give the average, because that doesn't provide very much information. An average height of 5'9" could mean that every male is exactly 5'9" tall, or it could mean that there are two groups in the population: 50 percent are 4'9" and 50 percent are 6'9". Or maybe 10 percent are 3'9", 25 percent are 4'9", 30 percent are 5'9", 25 percent are 6'9", and 10 percent are 7'9". The *average* height in each of these cases is 5'9", but a roomful of men from one of these populations would look very, very different from a roomful of another because the *distribution* of their heights is different. In a distribution where every male is 5'9" tall, there is zero probability that if you pull a male randomly off the street, he will be taller than

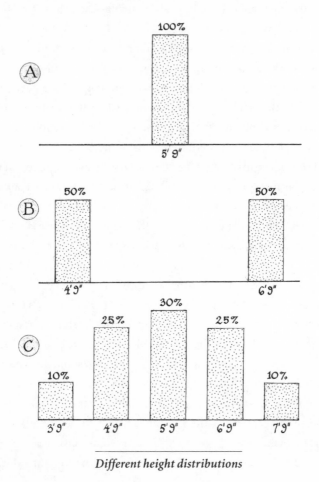

Different height distributions

6 feet. He must be 5 feet 9 inches exactly. But in the five-group distribution cited above, there's a 35 percent chance that if you pull someone off the street, he will be more than 6 feet tall (25 percent at 6'9" plus 10 percent at 7'9").

Of course, those examples don't really represent the true distribution of heights. In reality, the distribution of heights is very close to what is known as a *bell curve* distribution. In a bell curve, the "extreme" events are much, much rarer than "average" ones. For example, if you walk down the street, most adult men are within a few inches of 5'9" tall. It is rarer, but not uncommon, to see someone who is 5 inches taller at 6'2". You probably see dozens of them every day. But add another 5 inches and you will realize that it's pretty seldom that you run into someone who is 6'7"; depending on how many people you encounter, you might see one person like that in any given week. Add another 5 inches—7 feet tall—and you're in the stratosphere. You are not likely to run into very many seven-footers in a lifetime unless you're a basketball fan, and even in the NBA, seven-footers are worthy of note. This is a typical bell curve distribution; the probability of encountering a given event drops very rapidly as the event gets further away from the average and gets more extreme. Many everyday things—IQs, prices, shoe sizes—tend to follow a bell curve distribution.

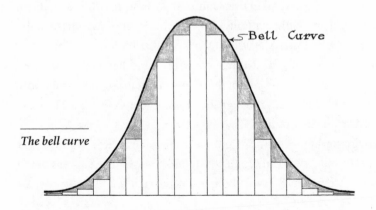

The bell curve

Maxwell's distribution had to do with atomic speeds. If you measured a random atom in a container of gas, what is the probability that it will have a given speed? It turns out that the resulting distribution is not quite a bell curve, though the two have some mathematics in common; it's more like a squashed and distorted bell curve, now known as the Maxwell-Boltzmann distribution.

Boltzmann got his name attached to the distribution because he proved, mathematically, that this was the distribution that gas atoms in equilibrium had to have. Maxwell showed that this particular distribution of speeds would fit the data, but Boltzmann proved that a set of billiard-ball-like atoms in a chamber—under certain basic assumptions—*must* have a Maxwell-Boltzmann distribution of speeds. Boltzmann's proof helped launch a stellar career in physics, but it led him into areas that made his ideas very unpopular with many physicists of the day.

For one thing, Boltzmann's proof wasn't based in experiment. It was based in pure mathematics. Instead of playing around with containers of gas and fitting his data to a mathematical function that *seemed* to explain the data, Boltzmann made a few simple assumptions, rearranged formulae, and *proved*, with 100 percent certainty, that if those assumptions are correct, the Maxwell-Boltzmann distribution was the *only* probability distribution that the gas atoms in equilibrium could take. Even more important, in 1872 Boltzmann proved, again mathematically, that in a container of gas whose molecules were not in a Maxwell-Boltzmann distribution (say, for example, you artificially filled a chamber with atoms that had precisely the same speed), the collisions of the atoms would make some atoms lose speed and some gain it, inevitably leading to a gas with a Maxwell-Boltzmann distribution. That is, start with a gas whose atoms are moving in any which way and let it settle for a little while; it will quickly and irreversibly come to equilibrium where the speeds of the atoms are in the Maxwell-Boltzmann distribution. This important scientific result wasn't based upon experiment or observation; it was based on pure deduction and was therefore considered a mathematical theorem rather than a physical law.[6]

But the mathematical nature of his work wasn't the biggest problem with Boltzmann's methods; after all, Newton's work was mathematical as well. What made Boltzmann different from Newton and all his predecessors was that Boltzmann's work dealt with probabilities and statistics—with distributions and random events and other unpredictable physical processes—whereas physics, from its very beginning, dealt only with certainties. If you knew a planet's position and its velocity, you knew precisely where it would be anytime in the next billion years. If you dropped a sphere off the Leaning Tower of Pisa, you knew, to within a fraction of a second, precisely when it would strike the ground. The immutable, firm laws of physics seemed to be the only certain things in the universe. By injecting probability and statistics into physics, it seemed as if Boltzmann was destroying the certainty of the beautiful, incontrovertible laws that governed Nature. Even the second law of thermodynamics.

In fact, Boltzmann didn't tear down the second law—he explained it and showed why the second law *must* be. But it didn't seem like it at the time. Boltzmann's work, which relied upon probability and randomness rather than certainty, seemed to undermine the very foundations of physical law; it seemed as if laws could only apply *some* of the time in Boltzmann's probabilistic and statistical universe. And at the center of this problem is a concept known as entropy.

You've probably already heard of entropy. In fact, it has already appeared in this chapter in disguise.[7] Most people think of entropy as

---

6. It came to be known as the H theorem, apparently because an English physicist mistook an ornate German capital letter *E* for an *H*.
7. "Equilibriumness" was a way of talking about entropy without having to formally introduce it.

a measure of disorder. If you ask high school physics teachers what entropy is, nine out of ten will describe it as a way of expressing how messy your bedroom is or how poorly you've arranged your books on a shelf. This is a valid definition, but it is deeply unsatisfying and somewhat misleading. After all, with a neat room or an alphabetized bookshelf, a human is arbitrarily deciding what *ordered* and *disordered* mean, when, in fact, entropy doesn't require anyone to judge what is neat and what is messy. Entropy is a fundamental property of a collection of objects; it comes from the laws of probability and from Boltzmann's statistical approach to physics. So, forget for the moment about order or disorder, messiness and neatness.

Imagine, instead, that in the middle of my living room I have a big box lying on the floor, its top gaping open. (This is true more often than I care to admit.) Imagine, too, that someone has painted a thin red stripe inside, dividing the box into two equal parts. Now, being the sort of person who has nothing better to do on weekends, I can amuse myself by tossing marbles randomly into the box. When I throw a marble, it has an equal chance of landing on either side. For any given marble that I throw, 50 percent of the time it will wind up in the right half of the box, and 50 percent of the time it will wind up in the left half. It's a pretty pathetic way to kill time—it's only slightly better than watching reality TV—but this simple setup is all we will need to understand the concept of entropy.

Let's start out with two different marbles. Plunk, plunk. And now let's look inside the box and see what happened.

When I peer into the box, I'm greeted by one of four possible outcomes. Case 1: The first marble I threw landed on the left side of the box, as did the second marble. Case 2: The first marble landed on the left, but the second landed on the right. Case 3: The first marble wound up on the right and the second settled on the left. Case 4: Both marbles landed on the right side of the box. Each one of these possibilities is equally probable; that is, there's a 25 percent chance for each case.

However, things change slightly if the marbles look exactly the same. In this case, you can't tell which one you tossed first, so when you look in the box, there are only three possibilities: both marbles are on the right; both marbles are on the left; or there's one on each side. In other words, case 2 and case 3 become indistinguishable (or *degenerate,* in physics-speak). This degeneracy means that the possibilities are no longer equally probable. As before, there's a 25 percent chance that both marbles are on the left side of the box and a 25 percent chance that both marbles are on the right side of the box. But the third possibility—that there's one marble on the right and one marble on the left—happens 50 percent of the time, because there are two ways it can come about. That means that having one marble on each side of the box is twice as probable as, say, having both on the left side of the box.

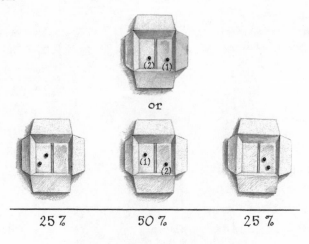

Two indistinguishable marbles in a box

Now let's scoop out the two marbles, and toss four in. Plunk, plunk, plunk, plunk. This time there are sixteen possible outcomes if we can keep track of each marble, but if the marbles are identical, we can only distinguish five different cases: (1) four marbles on the left and zero on the right, (2) three on the left and one on the right, (3) two on the left and two on the right, (4) one on the left and three on the right, and (5) zero on the left and four on the right. Don't worry too much about the probability calculations (you can see the details in the table below), but notice that the most probable outcome is six times as likely as the least probable outcomes. When you graph the probabilities—when you look at the probability distribution—you might notice that the probabilities are following that distribution so familiar to statisticians: a bell curve.

	Marbles on the left	Marbles on the right	Probability
Probabilities for four indistinguishable marbles in a box	4	0	$\frac{1}{16}$
	3	1	$\frac{4}{16} = \frac{1}{4}$
	2	2	$\frac{6}{16} = \frac{3}{8}$
	1	3	$\frac{4}{16} = \frac{1}{4}$
	0	4	$\frac{1}{16}$

Four indistinguishable marbles in a box

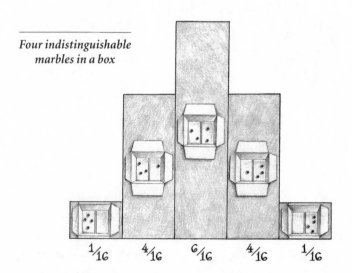

$\frac{1}{16}$ $\frac{4}{16}$ $\frac{6}{16}$ $\frac{4}{16}$ $\frac{1}{16}$

In fact, the more marbles we toss into the box, the clearer the bell curve becomes. No matter how many marbles we throw, on average half the marbles will fall into the left half of the box and half will fall into the right half, and this is always the most probable outcome of any given trial. The most extreme events are when all the marbles are on the right half, or all on the left half, and these extremes are much, much less probable than the average, or *mean,* event. All the other events lie in between the mean and the extreme and become dramatically less probable as they move from the mean to the extreme. And the more marbles we toss into the box, the less probable the extreme events become. For instance, let's throw a nice large sample of 1024 marbles into the box. On average, 512 will wind up on the left-hand side and 512 will wind up on the right. An extreme case, such as 1024 on the left and 0 on the right, is unimaginably improbable.

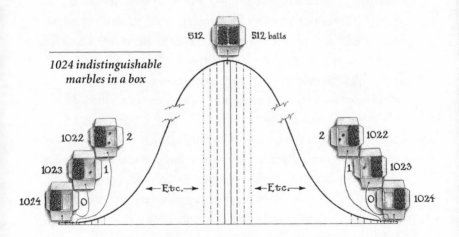

How improbable? Toss 1024 marbles randomly into a box. Look inside. Take out the marbles and toss them in again. Look inside. Take the marbles out and toss them in. Look again. Repeat and repeat and repeat. If you did this once a second from the beginning of the universe to now, the odds of ever seeing a 1024-on-one-side case are about

10^{290} to 1 against. Indeed, if every atom in the universe were one of these 1024-marble boxes, filling up randomly with marbles over and over again every second from the very beginning of the universe, not one of those boxes would ever have a 1024-on-one-side trial. (It's not even close. There are only about 10^{80} atoms in the visible universe.) While it is not *impossible* to randomly get all 1024 marbles on one side, it is so improbable that it's *functionally* impossible. It won't happen in this universe.

So what? Why waste our time playing around with boxes and marbles? Because it leads directly to a simple definition of entropy. In fact, entropy in this box-and-ball system is just a measure of the probability of any given configuration of marbles inside the box.

If you take a chunk of coal and weigh it, the number on the dial of the scale is a measure of how much stuff there is in that lump; given the weight and composition of any piece of matter, you can estimate how many atoms are in that lump, and of what sort they are. If you take a cup of coffee and stick a thermometer in it, the reading on the thermometer is a measure of how fast the molecules inside the coffee are moving. If you know the temperature of a chunk of matter, you know, roughly speaking, how those molecules are moving. Like mass and temperature, entropy is a measure of a property of a bunch of matter. If you know the entropy of a container full of atoms, you know, roughly speaking, how those atoms are distributed. And though entropy seems as if it is more abstract than temperature and mass, it is actually as concrete and as fundamental a property as temperature or mass.

Part of the reason entropy is harder to wrap your head around than mass or temperature is that what it's measuring is a bit harder to quantify than a table's mass or a bicycle's speed. Entropy captures the configuration of the entire collection of matter in terms of probabilities—in terms of the most probable configurations of a collection of atoms or, in our box-and-marble example, the most likely outcomes when we dump marbles into the box. The higher the probability of a configuration of matter (or the more likely an outcome of a box-and-marble experiment), the higher the entropy of that configuration (or outcome).

In the 1024-marble case the most likely outcomes—roughly 512 marbles on each side—have a high probability and a high entropy. The least likely outcomes—most of the marbles on one side or the other—have a low probability and a low entropy. In mathematical terms, if p is the probability of a given configuration, such as 512 marbles on each side, the entropy of that configuration, which physicists denote with the letter S, is just a function of $k \log p$, where k is a constant and "log" is a *logarithm*.[8] (The expression on Boltzmann's grave, $S = k \log W$, is equivalent.) If you shake up a container full of marbles and peer inside, the marbles will almost certainly be in a high-entropy configuration. That's it. That's entropy.

Seems simple enough. In fact, it even seems like a tautology. Toss marbles in a box; when you look inside, the marbles will probably be in a probable configuration. Well, duh. That's why we say that configuration is probable. But to physicists this observation is profound. The idea of entropy has deep and far-reaching consequences, and not just with marbles and boxes. Entropy is an unavoidable and troubling element of Boltzmann's statistical mechanics—and it is woven into the very nature of the universe.

A box full of marbles is very similar to a container full of gas, so let's lose our marbles. If we put 1024 atoms of helium, say, in an empty container and randomize it by shaking it up (in fact, the container shakes itself up because of the random motion of the atoms), whenever we peek in about half of the helium will be on the left side of the container and half will be on the right. The entropy will be high, and the atoms will be, more or less, uniformly distributed throughout the container. In fact, no matter what property of the atoms we look at, the highest-entropy state corresponds to a uniform distribution of that property. For example, whenever we peek into the container, the high-temperature, fast-moving atoms will tend to be evenly distrib-

---

8. Don't worry too much about this equation. I mention it so that you're familiar with its form, as it will crop up again later. See appendix A for a brief refresher on logarithms if you'd like to examine the equation in more detail.

uted throughout it; so will the low-temperature, slow-moving atoms. It is extremely unlikely that all the high-temperature atoms will cluster on the left side of the container and all the low-temperature atoms will clump together on the right. Instead, the gas will almost certainly be a uniform temperature throughout. That's the highest-entropy state, and it's a virtual certainty that the helium will be in that state when we peek in. Under conditions where an isolated container of gas is allowed to randomize—where it is allowed to evolve to equilibrium—we are almost guaranteed never to see the left side cold and the right side hot.

But what about a drafty room? It's cold near the window, while it's toasty near the radiator. At first glance, this might seem to contradict the concept of entropy. However, this is a system that is not isolated; the radiator keeps pumping warm air into the container, while the window lets it escape. If we were to seal off the window and turn off the radiator, the room would quickly reach an equilibrium where every place is the same temperature. Similarly, we can inject a bunch of helium atoms on one side of a container, knocking it out of equilibrium, but left alone, the container will rapidly revert from its low-entropy state (with lots of atoms on one side and few on the other) to its high-entropy state (a roughly equal number on each side). It's as if the system is attracted to its high-entropy state—and, in a sense, it is. Just as a ball "wants" to roll down a hill, a box full of gas "wants" to maximize its entropy. You can do work—put in energy—to reverse the trend toward high entropy in a system such as using an air conditioner or a heat pump to keep one side of the container hot and the other side cold, but left to its own devices, a container full of gas will revert to its maximum-entropy state, with hot and cold atoms evenly distributed throughout.[9]

Atoms' "desire" to maximize their entropy leads to an irreversible

9. Boltzmann's H theorem was, in fact, a theorem about entropy. When it comes to speed, the atoms maximize their entropy by assuming a somewhat distorted bell curve distribution of speeds: the Maxwell-Boltzmann distribution. But for clarity's sake, I'm going to ignore this distribution and just talk about "hot" and "cold" atoms, just as if I were talking about a billiard ball that was painted red or blue.

change in a container full of gas. If you start off with all the atoms in one corner of a box, then after a little while the atoms will spread out, maximizing their entropy. Since it's so improbable that all the atoms will move back to the corner they came from, the gas is essentially permanently in a state of high entropy: once it reaches equilibrium, the gas will always be in a high-probability state and will never revert to the low-probability state whence it came. Similarly, if you start off with a bunch of hot atoms on the right side of the box and cool atoms on the left side of the box, then after a while the hot atoms and the cool atoms will jostle around randomly and will move into the most probable configurations: the hot atoms and cold atoms will be equally distributed on the right and left sides of the box. And once the box is in equilibrium, you can watch it for centuries and you will never see the evenly distributed hot and cold atoms suddenly separate themselves and wind up on opposite ends of the box. If left to itself—if you don't use a heat pump or add any energy to the system—the increase in entropy is irreversible.

This irreversibility is a key feature of entropy. On the microscopic scale, the atoms are smacking into one another like billiard balls, ricocheting off one another and slamming into walls. If someone showed you a short film of two moving billiard balls bouncing off each other, you might have a difficult time telling whether the film was being played forward or backward. In both cases, the balls slam into each other and fly away; in neither case does the perceived motion violate a law of physics. The collision of the atoms is *reversible;* the reversed collision is just as valid and likely as the forward collision. But even though each individual action of those atoms is reversible, the *collective* motion of the atoms is irreversible. If you saw a movie where all the atoms in a container gathered themselves into one corner, you would know that it was being played backward. You would realize instantly that the original film showed the gas being released from one corner and spreading out from there; in real life, gases always behave in this way, but not in the reverse direction. Because of entropy, one direction is allowed by the laws of physics and the other is (essentially) forbidden. Entropy makes the behavior of the gas irreversible; because of entropy, the film only makes sense when played forward, not backward. You can't reverse the direction of the movie. For this reason, scientists often refer to entropy as "the arrow of time." The irreversibility of entropy-changing reactions is a signpost that tells you which way time flows. Time goes forward as entropy increases; it never goes backward because, in a system left to its own devices, entropy doesn't decrease.

Entropy is also the key to understanding thermodynamics. In a sense, heat engines are simply machines that increase the entropy of the universe. When they pump heat from a hot reservoir to a cold reservoir, they are increasing the entropy of the system as a whole. The separation of the hot reservoir from the cold reservoir is, inherently, a low-entropy configuration, just as a box with hot atoms on one side and cold atoms on the other is low entropy as well. By allowing heat from the hot reservoir to flow into the cold reservoir, you're getting the system closer to equilibrium; you're increasing the system's entropy. And the system's "desire" to increase its entropy is so great that you can put a device in between the two reservoirs and make it do work for you.

Conversely, you cannot reverse the increase in entropy without doing work; you have to add energy to the system to reverse its trend toward equilibrium, and in doing so, you have got to increase the entropy of the universe outside your system even more than you are reducing the entropy of the gas inside your system. This is the essence of the second law of thermodynamics: entropy is supreme. The universe is slouching toward a state of higher entropy, and there is nothing you can do to reverse it. You can impose order on a little corner of your universe—you can set up a refrigerator that separates heat from cold in your kitchen—but you must consume energy to do so, and that consumption of energy increases the entropy of the Earth more than the refrigerator decreases the entropy of your kitchen. It's a disturbing idea: you are bringing the Earth infinitesimally closer to a state of chaos when you chill down a bottle of beer in your fridge.

Boltzmann's statistical and probabilistic view of the motion of atoms in matter was incredibly powerful. By looking at a gas as a collection of randomly moving particles, he was able to explain the physical principles that drove engines, that were responsible for heat flow, for temperature, for work—and for entropy. Most of all, for entropy. Through simple probability and statistics, Boltzmann's work led to the understanding that systems naturally "try" to increase their entropy, and that the universe as a whole is constantly, irreversibly, getting more entropic. But hidden in his logic was a time bomb.

The probabilistic nature of Boltzmann's work made it appear as if it undermined the absolute truth of the very laws it explained. The second law of thermodynamics was based upon the fact that gases *probably* wind up in their most probable configurations. Sounds redundant . . . but probably isn't absolutely. Once in a while, perhaps, the gas randomly winds up in an improbable configuration—it can happen. That means that the entropy of the system, without any energy being added, can spontaneously decrease. The second law, to all appearances, is suddenly violated. Worse yet, James Clerk Maxwell, the man who embraced the statistical nature of gases and came up with the distribution of speeds of the atoms in a gas, devised a clever method that seemed to separate hot atoms from cold atoms without any work at all—an even more profound violation of the second law.

Boltzmann proved that the second law must be true. But at the same time, his methods apparently undermined the law and showed that it need not be true all the time. This was a demon that followed Boltzmann throughout his career.

In 2002, a group of Australian scientists published an article in the journal *Physical Review Letters* that caused a minor ruckus. No wonder, as its title was very provocative: "Experimental Demonstration of Violations of the Second Law of Thermodynamics for Small Systems and Short Time Scales."

The scientists—who were from the Australian National University in Canberra and Griffith University in Brisbane—performed a clever measurement of the entropy of tiny latex beads in water. Like atoms in a box, these beads—about one hundred at a time—floated around randomly in a container of water. Using a laser, the researchers trapped the minuscule spheres and released them, and then measured how the entropy of the system evolved.

Most of the time, the beads behaved exactly as you'd expect: the order that the laser imposed quickly disappeared, and the system's entropy increased. But once in a while the entropy decreased slightly for a short time before it increased again. For a small time, in a small system, entropy spontaneously decreased rather than increased. Hence the "violations" of the second law. If even for a small time, the second law seemed to fail. Entropy was going down, not up.

As you would expect, the news media played up the apparent breach of the most fundamental law in physics. But though the 2002 experiment (and a more carefully done follow-up in 2004) showed that entropy did, in fact, decrease for a short time in a small system, it wasn't a violation of the second law. It was exactly what the statistical nature of the law allows. So it wasn't nearly as big a deal as the media portrayed it.

When Boltzmann formulated his picture of gases in boxes, he knew that a large group of particles, even though they individually move in a random manner, collectively are predictable. The more particles there are in a system—the bigger it is—the firmer the predictions become.[10] But conversely, the smaller the system is, the more susceptible the prediction is to a random fluctuation.

Boltzmann painted the second law of thermodynamics as a probabilistic law. It holds true with statistical certainty. In reasonably large systems, you won't see a violation of the law at any point in the entire lifetime of the universe. (Remember, in the 1024-marble case, you would never see all 1024 marbles on one side of a box even if every atom in

---

10. This is a mathematical principle known as the *law of large numbers*. In essence, it says that the size of a deviation from expected behavior gets smaller and smaller as the number of random events gets larger and larger.

the universe were a box full of marbles, each box randomly refilling over and over again every second from the birth of the universe until now.) But in small systems, such as the one with four marbles, you will occasionally see all four wind up on one side of the box or the other. (In fact, it happens one-eighth of the time.) So, if you have a box with four marbles in it in a maximum state of entropy—two on each side—and you shake it, there's a one in eight chance that it will spontaneously decrease its entropy to the minimum possible. Though this reduction in entropy seems like a violation of the second law, it really isn't. This sort of thing is simply a consequence of the statistical nature of the law.

Modern physicists realize that even the most solid laws—even the second law—have a statistical element to them. For example, for a short time and on very small distance scales, particles can wink in and out of existence, owing to what are known as *vacuum fluctuations*. No physicist sees that as a true violation of the law of conservation of mass and energy. These fluctuations are something that modern physicists have come to terms with. But in Boltzmann's day, the lack of the absolute, cast-iron requirement for entropy to *always* increase was a major strike against his theory. But an even more serious challenge to Boltzmann came from the man who inspired him: Maxwell.

Boltzmann loved Maxwell's work, and Maxwell's 1866 paper on gases led to Boltzmann's work on the speed of atoms. Boltzmann likened the 1866 paper to a symphony:

First, the variations in velocity develop majestically, then the equations of state enter on one side, the equations of motion on the other; ever higher surges the chaos of formulas. Suddenly, four words sound out: "Put N = 5." The evil demon V vanishes, just as in music a disruptive figure in the bass abruptly falls silent.[11]

11. Quoted in Lindley, *Boltzmann's Atom*, 71.

Maxwell would soon summon a demon rather than dispel one. In 1871, he published *Theory of Heat,* in which he attempted to poke a hole in the second law of thermodynamics. Maxwell came up with an ingenious way to exploit the random motion of atoms to reverse the ravages of entropy and create a perpetual motion machine—he thought he had found a major flaw in the second law.

The scheme involved a tiny intelligent "being" in a box full of gas. The box has a wall, dividing the container into two equal halves. Embedded in the wall is a frictionless shutter. By opening and closing this shutter—an act that, thanks to the lack of friction, requires no work to perform—the tiny being can either let an atom pass from one side of the box or refuse it passage. Maxwell realized that this tiny being—which the physicist William Thomson, Lord Kelvin, would soon dub "Maxwell's demon"—could systematically reverse entropy without seeming to consume any energy or do any work.[12]

Maxwell's demon

12. Ever the Englishman, Thomson replaced the single demon and sliding shutter with a legion of demons wielding cricket bats.

For example, the demon could start off with a box in a state of high entropy—the hot and cold atoms evenly mixed throughout the box—and end up with all the hot atoms on the left and the cold atoms on the right. All the demon would have to do is open and close the shutter at the appropriate times. If a cold atom on the left side of the box approaches the shutter, the demon would let it through, but it wouldn't let any hot atoms pass from left to right. Conversely, it would open the shutter if a hot atom was moving from right to left but slam it shut if a cold atom was about to escape from its confinement on the right.

After time, with seemingly no work expended, the demon could segregate the box into a hot zone and a cold zone—a state with much, much less entropy than the original equilibrium of the box. The demon simply exploited the random motion of the molecules and let them sort themselves.

This was a much more serious challenge to Boltzmann's work than the mere objections to the statistical nature of his laws. It seemed that a properly designed piece of machinery might spontaneously reverse the entropy in a box, creating a hot reservoir and a cold reservoir without expending any energy. If this were possible, then you could hook up a Maxwell's demon to a heat engine; the engine would produce work while the demon kept the hot reservoir hot and the cold reservoir cool. You would get work for free—a perpetual motion machine.

Sadly, Boltzmann didn't live to help overcome Maxwell's demon; he succumbed to the struggle with his own. Boltzmann was often prickly and antisocial, and his novel ideas made him powerful enemies. On top of that, he was prone to bouts of depression and exhaustion. He hanged himself never knowing the secret that would lead physics to victory over Maxwell's demon. Ironically, the formula at the center of that victory was inscribed on Boltzmann's grave: $S = k \log W$, the formula for the entropy of a container full of gas. But it wasn't entropy that defeated Maxwell's demon. It was information.

CHAPTER 3

INFORMATION

—What do you want?
—Information.
—You won't get it.
—By hook or by crook, we will.

—The Prisoner (television series)

The concept of information itself was not new. But in 1948, when an engineer-mathematician realized that information could be measured and quantified—and that it was intimately linked to thermodynamics—he sparked a revolution and killed a demon.

The theory of information did not seem all that important at first. True, it changed the way cryptographers and engineers thought about their work; true, it set the groundwork for building the computers that would soon become part of everyday life. But even the founder of information theory, Claude Shannon, had no idea just how far-reaching his idea would become.

Information is much more than the redundancy in a general's code or, later, the ons and offs of computer switches. Though it can be represented in many ways—by the pattern of ink on paper, by the flow of electrons through a circuit board, by the orientations of atoms in a piece of magnetic tape, or by lights that are flipped on or off—there is something about information that transcends the medium it is stored in. It is a physical entity, a property of objects akin to energy or work or mass. Indeed, it would become so important that scientists would soon learn to recast other theories in terms of the exchange or manipulation of information. Some of the most fundamental rules in physics—the laws of thermodynamics, for example, and the laws that tell how collections of atoms move in a chunk of matter—are, deep down, actually laws about information. It is by looking at Maxwell's demon in information-theoretic terms that scientists were able to dispel it at long last.

Information theory would take the equation from Boltzmann's grave and use it to slay the demon that threatened to tear down the edifice of thermodynamics. Nature seems to speak in the language of information, and when scientists started to understand that language, they began to tap into a power that even Shannon never imagined.

The hero of information theory is Claude Elwood Shannon, who was born in 1916 in Michigan; as a boy, he was always something of a tinkerer, so it was natural that he wound up studying engineering and mathematics. These two disciplines intertwined throughout his life—and wound up converging in the theory of information that he would later create. In the 1930s, Shannon was bridging the two disciplines by working on a machine to solve a specific type of mathematical construct called a differential equation.

A run-of-the-mill equation, like $5x = 10$, is really a question of a sort: What number, when plugged into the place of x, will satisfy the expression? Differential equations are similar, but the questions are a bit more intricate, and the answers are themselves equations, not numbers. For example, a physics student might plug in the dimensions and other properties of a metal bar, as well as the temperature of a flame at one end, into a differential equation; out would pop an equation that explains how hot any given part of the bar is at a given time. These equations are fundamental to physics, and scientists at the time were desperately trying to find ways of solving them quickly with primitive computers. Shortly after graduating college, Shannon got a part-time job at the Massachusetts Institute of Technology (MIT), where he worked on a mechanical differential-equation solver that had been developed by Vannevar Bush, a scientist who, within a decade, would become one of the important people behind the development of the atomic bomb. Shannon helped translate differential equations into a form that the computer could understand, and eventually he started thinking about the designs of the electrical relays and flip-flop switches that sat at the core of the differential-equation computer. His master's thesis, written while he was working part-time at MIT, showed how engineers could use Boolean logic—the mathematics of manipulating *1*s and *0*s—to design better switches for electrical equipment (including computers).

After he finished his PhD, Shannon landed at Bell Laboratories. As the name would imply, Bell Labs was the research arm of the American Telephone and Telegraph Company (AT&T), the corporation that had a monopolistic hold on the United States' telephone system. The lab, founded in the 1920s, was intended to do basic research that was relevant to communications. The scientists and engineers there helped pave the way for high-quality sound recordings, television transmissions, advanced telephony, fiber optics, and other mainstays of the way our society communicates. At its core, communication is simply the transmission of information from one person to another, so it should come as no surprise that the lab's work wound up in areas that you'd think of as "information technology." The first binary digital computer and the transistor, for example, were developed at the lab.

Shannon was perfectly suited for work at Bell Labs, and he soon embarked on a project that would change the world of science. At first glance, it wouldn't seem as if his research could be so revolutionary. It dealt with how much capacity a given telephone line (or radio connection or any other communications "channel") can have. It's a very nuts-and-bolts question; the engineers at Bell Labs wanted to know how to pack as many different telephone conversations as possible on the same line at the same time without having the calls interfere with one another. In other words, how can you pack the most information possible on a single copper cable?

Communications scientists were in uncharted territory. Engineers from Roman times knew basic principles for building bridges and roads; even the science of thermodynamics was about a century old. But telephony was something entirely new. A bridge builder who wants to find out how much traffic a bridge can carry can calculate how much each car will likely weigh and how strong the steel beams that support the traffic need to be. He can use the concept of mass to figure out the capacity of any given bridge. But doing the same thing for a telephone line left engineers totally in the dark. There was no obvious way to calculate how many calls a company could cram on a single telephone line at the same time. Just as bridge builders needed to be able to understand and measure mass to figure out the capacity of a bridge, engineers had to learn to understand and measure information to figure out the capacity of a telephone line. Shannon was the one who provided this basic understanding, and it had much greater repercussions than merely helping Ma Bell.

When Shannon set out to answer the question of telephone line capacity, he put together all the elements of mathematics and engineering—the knowledge about the nature of questions and answers, about machines, about Boolean logic, and about electrical circuits. When he did, he created the third great revolution in physics in the twentieth century: as did relativity and quantum theory, information theory radically changed the way scientists look at the universe. But Shannon's information theory started off small and in familiar territory: in the realm of questions and answers.

Shannon's first great insight came when he started thinking about information as something that helps you answer a question: What is the solution to this differential equation? What is the capital of Burkina Faso? What are the component particles that make up the atom? Without the proper information, you are unable to answer these questions. Perhaps, based on the limited knowledge—information—in your head, you can make a few uncertain guesses. But even if you don't know the answer right now, you can figure it out with great confidence if someone sends you the proper information.

So far, this is quite abstract, so let's take a concrete example. On April 18, 1775, just before the outbreak of the American Revolution, the Americans knew that British troops were about to make a move. They knew that the British army, which was gathered in Boston, was likely to march north to Lexington, but there were two possible routes that the army could take. The first way was simple but lengthy: the army could march southwest from Boston through a narrow strip of land and then veer north toward their target. The second route was more logistically difficult, but quicker: the army could ferry across the mouth of the Charles River and march immediately northward to Lexington. The question was, Which route would the British take?

There were two possible answers to this question: by land or by sea. Patriots waiting on the north bank of the Charles River had no information about the British strategy, so they had no idea where to organize their defense. As soon as the British started to move, everybody in Boston would immediately know the route that the redcoats would take, but that information would not be available to the minutemen in Lexington. Until somebody sent them the answer to the question—the information about which way the British were marching—the Americans couldn't begin their defense.

Luckily, about a week earlier Paul Revere and a number of other American patriots had set up a scheme for gathering and transmitting that information to the defending troops. As soon as the British would begin to move, the sexton of the Old North Church in Boston—like all other Boston citizens—could see which route the British were going to take. The sexton would then climb the steeple of the church and hang lamps to communicate the route to the Americans on the other shore. One lamp would mean that the British were taking the long, landbound route; two lamps would mean that the British were ferrying across in boats. One if by land, two if by sea.

When two lamps appeared in the steeple that evening, the patriots instantly knew the answer to the question. The information in that message took away any uncertainty about the British plan; the patriots knew, for certain, that the British were coming by ferry and that they would arrive soon. (Of course, any lingering doubt was soon dispelled by the clatter of Paul Revere riding through town, broadcasting it directly.)

From Shannon's point of view, this is a classic example of the transmission of information. Before the message—before the lamps were hung in the steeple—the recipients of the message, the American patriots, could only guess, and any guess had a 50 percent chance of being wrong. But once the lamps were hung, the message was broadcast, and information was transferred from the church's sexton to the American patriots. Two lamps answered the patriots' question; there was no longer any uncertainty about the route the British were taking. Now they were 100 percent certain about which route the British army would use. The message reduced the Americans' uncertainty—in this case, to zero—about the answer to the question, and that, to Shannon, is the essence of information.

But the real power of Shannon's view of information is that it gives a measure of *how much* information is transmitted in a given message. He realized that a simple question like this—which has two possible answers—was essentially a yes/no question. Are the British coming by land or by sea? Are you male or female? Is the coin flip heads or tails? Is a light on or off? All of these can be rephrased in simple yes/no terms. Are the British coming by sea? Are you female? Is the coin flip heads? Is the light on? In each case, a no answer still leaves no uncertainty about the answer to the question. If the British aren't coming by sea, they're coming by land. If you're not female, you're male. If the coin isn't heads, it's tails. If the light isn't on, it's off. Thus a yes/no question suffices for each of these queries. And mathematics has a great way of dealing with yes/no questions: Boolean logic.

Boolean logic deals in trues and falses, yeses and nos, ons and offs.

The answer to any of these simple yes/no questions can be represented by a single symbol from a matched set of two: T vs. F; Y vs. N; 1 vs. 0. Take your pick. (For the sake of consistency in this book, I will use 1 for a "true/yes/on" answer and 0 for a "false/no/off" one.) Question: Are the British coming by sea? Answer: 1. Question: Is Tony Blair female? Answer: 0. A yes/no question can always be answered by a single symbol that can take one of two values. This symbol is a *binary digit*, or bit.

The term *bit* first appeared in Shannon's 1948 paper "A Mathematical Theory of Communication," which set out the foundation of what is now known as information theory.[1] In Shannon's theory, a bit became the fundamental unit of information.

Answering a yes/no question requires one bit of information. You need to set up a binary digit in the steeple of the Old North Church to distinguish whether the redcoated Brits are coming by land or by sea; a 0 means land and a 1 means sea. Transmit that digit in a message and you answer the question. But it doesn't matter at all what *form* this message takes. It could be one lamp versus two lamps in the steeple, or perhaps a red light versus a green light. It could be a flag on the left side of the church versus a flag on the right side, or a rumble of a cannon shot into the air versus the lighter crack of a volley of musketry fire. Even though the media are all different, the information in the message is the same. No matter what form the message takes, it carries one bit of information, allowing the American patriots to distinguish between the two possibilities and answering the question of which route the redcoats would take.

But what happens if the question is more complicated and cannot be answered with a simple yes or no? For example, what if the British could also take a train from Boston and wind up at the Lexington sta-

---

1. Shannon credits his Bell Labs colleague John Tukey with coining the word; thankfully, *bit* replaced the much uglier *bigit*, which was beginning to circulate at the time. Later wags would coin the terms *byte* for eight bits and *nibble* for four bits—half a byte. (Tukey, incidentally, would be known for codeveloping one of the most important algorithms in computer science, the fast Fourier transform, but that's another story altogether.)

tion? Or if they could fly in, parachuting from eighteenth-century hot-air balloons directly into the Massachusetts town? With four possibilities, no longer can a single bit of information completely answer the question of how the redcoats are coming.

In this case, before the message is transmitted, the American patriots have four possibilities to choose from, each of which is presumably equally probable. The patriots could make a guess, but absent any information they would only guess right 25 percent of the time. And the one-bit message answering the question, "Are the British coming by sea?" will only reveal the answer one-fourth of the time; a **0** answer to that question—one light in the Old North Church—still leaves it ambiguous whether the British are coming by land, train, or air. The message "not sea" does not completely answer the question; one bit is not enough.

Paul Revere would have had to come up with a different scheme to answer the question completely; he would have had to come up with a way of transmitting more than one bit of information. For example, he might hang up to four lamps in the steeple: one if by land, two if by sea, three if by train, and four if by parachute. If there were eight possibilities, he might hang up to eight lamps in the church: one if by land, two if by sea, three if by train, four if by air, five if by hovercraft, six if by spaceship, seven if by teleportation, and eight if on the backs of a band of evil angels. That's a lot of lamps to cram into a steeple.

But if Revere were *really* smart, he could alter his scheme slightly to reduce the number of lamps required. Instead of having up to four lamps to distinguish among four possibilities, the message sender can use only two. Attach a filter to each so that it glows either red or green in the steeple and you can use them to tell you which way the British are coming: red–red means by land; red–green means by sea; green–red means by train; and green–green means by air. Two lights that can be either red or green—two bits—completely answer a question if there are four possible answers. You need two bits of information to distinguish among four scenarios. Similarly, three red/green lights, three bits, can answer a question if there are eight possible answers. You need three bits of information to distinguish among eight possibilities.

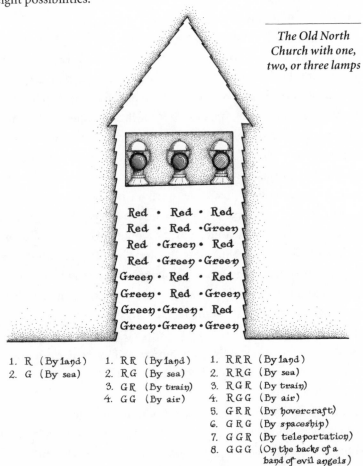

The Old North Church with one, two, or three lamps

Red · Red · Red
Red · Red · Green
Red · Green · Red
Red · Green · Green
Green · Red · Red
Green · Red · Green
Green · Green · Red
Green · Green · Green

1. R (By land)	1. R R (By land)	1. R R R (By land)
2. G (By sea)	2. R G (By sea)	2. R R G (By sea)
	3. G R (By train)	3. R G R (By train)
	4. G G (By air)	4. R G G (By air)
		5. G R R (By hovercraft)
		6. G R G (By spaceship)
		7. G G R (By teleportation)
		8. G G G (On the backs of a band of evil angels)

No matter how complicated a question is, no matter how many (finite) possible answers a question might have, you can answer the question with a series of bits, a series of answers to yes/no questions. For example, if I tell you that I am thinking of a number between 1 and 1000, you can figure it out by asking me only ten yes/no questions. Is it bigger than 500? No? Is it bigger than 250? No . . . and so forth. By the tenth question, if you've asked your questions right, you're guaranteed to know the answer with 100 percent certainty.

If, at the beginning of the game, you simply guessed what number I was thinking of, you would only have a 1/1000 shot—a 0.1 percent chance—of being correct. But each yes/no question that I answer about the number is giving you one bit of information, reducing your uncertainty further. Is it bigger than 500? No. This means that the number must be somewhere in the range of 1 to 500; there are only 500 possibilities, not 1000. If you now were to guess at the number, you would have a 1/500 chance of being right. Still not good odds, but twice as good as before. Is it bigger than 250? No. Now you know that the number is in the range of 1 to 250; there are now only 250 possibilities, and you have a 1/250 shot of being right if you guess. After three questions, you have a 1/125 shot of guessing correctly; after seven questions, about a 1 in 8 shot—about a 12 percent chance—of being right. After ten questions, you know the answer with 100 percent certainty. Each yes/no question reduces your uncertainty about the answer to the question of what number I am thinking of; each response to one of your yes/no queries gives you one bit of information. Distinguishing among 1000 possibilities requires only ten bits; with ten bits of information, a string of ten **1**s and **0**s, you can, with 100 percent certainty, answer a question with 1000 possible outcomes.

Shannon realized that a question with N possible outcomes can be answered with a string of $\log N$ bits—you need only $\log N$ bits of information to distinguish among N possibilities.[2] So, to distinguish between two outcomes, you need one bit; four outcomes, two bits; eight outcomes, three bits; and so forth. This principle has enormous power. I could tell you that I have picked out an atom somewhere in the universe. Since there are only 10^{80} atoms in the universe, and $\log 10^{80}$ is about 266, it would only take 266 properly chosen yes/no

2. In this case, the *log* symbol represents the logarithm base 2. That is, $x = \log N$ is the solution to the equation $N = 2^x$. Mathematicians often ignore the base of the logarithm; see appendix A on logarithms to find out why.

questions and 266 bits of information to figure out which atom I am thinking of!

However, information isn't only about guessing numbers and answering yes/no queries; it wouldn't be terribly helpful if it were just useful for winning games of twenty questions. Information—encoded in 1s and 0s and measured in bits—can be used to convey the answer to *any* question, so long as that question has a finite answer. This is true even for more open-ended questions, those that aren't obviously answerable by a set of yes/no questions, such as, What is the capital of Burkina Faso? If you asked me that question, I would have to communicate the answer to you somehow, and it's hard to imagine coming up with a string of yes/no questions, a stream of bits, that yields the answer, Ouagadougou. In fact, though, this is precisely how I'm answering this question as I type this manuscript into my computer. My word processor has encoded the stream of English letters that spell out "Ouagadougou" into a series of bits, a set of 1s and 0s on my hard drive. It does this by changing the symbols that make up the English alphabet into 1s and 0s—in essence, a series of answers of yes/no questions that spell out "Ouagadougou" on my computer screen. Since the English alphabet has only twenty-six characters, you theoretically need (a little less than) five bits to encode each letter. Since "Ouagadougou" has eleven letters, then eleven strings of five bits each would suffice to spell out the name—fifty-five bits completely answer the question, What is the capital of Burkina Faso?[3] These bits are stored on my hard drive, then transmitted to my editor in an e-mail. My editor's e-mail reader and word processor translate those bits back into written language and print them in a format that you and I can understand. It's a tortuous journey, but fundamentally I have

---

3. In actuality, computers tend to represent letters with more than five bits. One very common scheme, ASCII, encodes each letter with a byte of information—eight bits. This is more than you need to encode the English alphabet, but it gives you room for lowercase and capital letters, punctuation marks, foreign letters, and a number of other useful symbols.

answered the question, What is the capital of Burkina Faso? with a stream of bits—answers to a set of yes/no questions—that, taken together, give the correct answer.

Written language is just a stream of symbols, and symbols can be written as a stream of bits. So any question that has an answer that can be written in a language, any question that has a finite answer of any sort, can be written as a stream of **1**s and **0**s. What's more, Shannon realized that any question whose solution can be expressed in a finite way could be *answered* with a string of bits. In other words, any information, any answer to any finite question, can be expressed in a series of **1**s and **0**s. Bits are the universal medium of information.

This is a stunning realization. If any information, any answer to one of these questions, can be encoded as a string of bits, then it gives you a way to measure how much information there is in a message. What's the minimum number of bits you need to encode the message? Fifty bits? One hundred bits? One thousand bits? Well, that's precisely how much information the message contains. That's the measure of information in a message: how many bits you need to transmit it from a sender to a receiver.

Shannon also saw that the reverse logic held as well. If you intercept a message, if you grab a stream of symbols, such as letters of an alphabet, you can estimate the maximum amount of information that the stream can contain—even if you don't know the nature of that information. This leads to some rather creepy analyses. A typical 70,000-word book, such as this one, contains about 350,000 letters. Each of these can be encoded in five bits, so all told, a book like this can contain less than two million bits of information, and it usually contains a lot less than that. (More on this shortly.) Two million bits is about 0.25 percent of the capacity of a typical CD disk, or 0.04 percent of the capacity of a DVD. So, in information-theoretic terms, this book can carry as much information as about eleven seconds of the latest Britney Spears album or two and a half seconds of the movie *Dumb and Dumber.*

Of course, this analysis doesn't tell you how much information these media actually carry—it tells you the *maximum* amount that they can. It also doesn't tell you the nature of that information. It takes a lot more information to tell a TV screen how to paint tens of pictures per second or make a speaker warble in just the right way than it does to arrange a string of squiggles on a piece of paper. Not all of the information on a CD or a DVD is answering questions that are noticeable by humans, but it is information nonetheless. Is pixel number 3140 black or deep, deep brown in frame number 12,331 of *Dumb and Dumber*? Does Britney screech an E-flat for 3.214 or 3.215 seconds? We might not notice, or care about, the answers to these questions, but the CDs and DVDs answer them all the time, and that requires a lot of information. That's why a CD is required to store so much information, and why a DVD can carry even more. By comparison, a book is an information desert. Making matters even more depressing for a book author, a stream of written characters in a human language carries much less information than the maximum that a stream of twenty-six symbols can.

Before exploring the information content of language, let's return to a very simple case: a string of binary digits. As we saw, each digit in the stream can, potentially, carry one bit of information. But this isn't always the case. Imagine that someone sends you a string of 1000 bits—a message that could contain 1000 bits of information—perhaps a paragraph's worth of text encoded in binary. But when you get the message, you are surprised to see: **1111111111**. . . . Intuitively, you can see that this string doesn't contain very much information at all, and in fact, in information-theoretic terms it probably doesn't, either.

I didn't give you the whole string. In fact, I just gave you ten of the **1**s and you were able to deduce that the rest of this 1000-bit string was also made up of **1**s. I gave you a mere 1 percent of the digits, and you could, without thinking very hard at all, generate the remaining 99 percent. So, in a mere 10 bits, I was able to send you the entire message—and I could probably do it in fewer. If I said the string was

1111 ... or **11** ... or even **1** ..., you would have been able to figure out what the *entire* message was. In other words, I've compressed a 1000-digit message into a single binary digit; one bit was enough to tell you what the whole message was. But if the message can be compressed into a single bit, it must only be able to carry a single bit of information or less.

Similarly, the message **010101** ... can be compressed into about two bits; it probably has at most two bits of information in it. And the message **0110011001100110** ... has only about four bits, even though the full string of 1000 digits can, in theory, contain much, much more information. These strings are compressible if they are predictable. You can come up with a few simple rules that will generate the entire string of digits. And if a digit gets messed up in transmission—maybe the 750th digit in the string **11111** ... is a **0** rather than a **1**—those rules let you know that the **0** is probably an error. The rules that allow you to generate the entire message from just a few bits allow you to correct the string if someone makes a typo. The rules make the string *redundant*.

So we've come full circle. The first chapter introduced information as what is left when you remove all the redundancy from a string of symbols. This chapter started with a formal definition of information and redundancy followed from it—and though we haven't *formally* defined information-theoretic redundancy, it is precisely what the first chapter was referring to. Redundancy is the extra stuff in a string of symbols, the predictable part that allows you to fill in missing information. Because of unwritten rules, patterns in the string of symbols, we can ignore much of the message and even remove parts of it. In the string **11111** ..., we can get rid of nearly all of the digits and still reconstruct the entire message. That's because the message is simple and highly redundant.

Computer scientists are well aware of the redundancy in a stream of bits and bytes for two main reasons. The first is error correction. Humans make mistakes when entering long strings of numbers, so credit cards, serial numbers, bar codes, and numerous other numbers are padded with redundancy so that a computer will be able to detect whether someone has made a data-entry error.[4] But even more important, like humans, computers aren't infallible. CPUs (central processing units) make errors when multiplying or adding; memory addresses accidentally flip bits or fail completely; hard drives lose data. Computers need to be accurate despite these errors, so there is some redundancy built in to computer protocols. A computer can use them to detect and correct any errors that it makes. (Error correction is absolutely crucial to the operation of computers.)

The second reason that computer scientists are aware of redundancy is that computer files are nothing more than 1s and 0s written in the magnetic coating on a hard drive or inscribed upon a similar storage device, so by removing the redundancy and leaving the information, engineers can compress a computer file to make it take up less space on your disk. A text file on my hard drive—the initial chapter of my first book, *Zero*—contains 581 words and takes up about 27,500 bits of space. After squashing it with a commercial compression program, it takes up only about 14,000 bits and still contains the same amount of information.

It should come as no surprise that a text file can be drastically squashed without losing information. We already looked at how English and other human languages have a great deal of redundancy built into them. The unwritten rules behind grammar and spelling and proper English usage all impart a great deal of redundancy into

---

4. In fact, look at the beginning of this book. On the page with the copyright information, there is an ISBN, a code that has built-in redundancy; the last digit/letter is a *check* to make sure that the others have been entered correctly. For the really curious and geeky, here's how the ISBN code works: ignore the check digit for the moment—the last one set off by a dash—and then multiply the first digit by 10, the second digit by 9, and so forth, until you've multiplied the ninth digit by 2. Add them all together, divide that sum by 11, and take the remainder. Subtract that remainder from 11 and that's your check digit; in the case that your answer is 10, the check digit is the symbol X. Of course, there's also a bar code on the back, which also has a built-in check, but that's another story.

English; given an incomplete stream of English letters, we can often complete the stream without too much effort. English letters are symbols like any other, so written English—a stream of these symbols—is no different in principle from a stream of **1**s and **0**s. Like any highly redundant string of symbols, English can be severely compressed without losing any information.[5]

This compression is actually pretty severe. Even though you need five bits to specify a character in a stream of text—more if you make the distinction between uppercase and lowercase letters—it turns out that each English letter carries only, on average, between one and two bits of information.

One of the great victories of Shannon's information theory is in formally defining redundancy and figuring out precisely how much information can be carried in a stream of symbols—redundant or otherwise. This became Shannon's famous channel capacity theorem. It was originally intended to help engineers figure out how much stuff could be sent over a communications channel (such as how many phone calls a given telephone line can handle), but it wound up changing forever the way scientists looked at information. This theorem gets its power because Shannon analyzed information sources with a surprising tool: entropy.

The central idea in Shannon's information theory is entropy. Entropy and information are closely tied to each other; entropy is, in fact, a measure of information.

One of the central ideas that led to the channel capacity theorem was Shannon's derivation of a mathematical method for measuring information. In 1948, he came up with a function that allowed him to analyze

---

5. It's not only written language that's redundant. Spoken language is also a stream of symbols, though the symbols are auditory rather than written. The basic symbol of spoken language is the *phoneme* rather than the letter, but once you take that into account, the same analysis applies. One of the great strengths of Shannon's theory is that it really doesn't matter *how* the information is conveyed; the mathematics stays the same.

the information in a message or sent over a communications line in terms of bits. In fact, Shannon's function looked exactly the same as the one Boltzmann used to analyze the entropy of a container full of gas.

At first, Shannon was uncertain what to call this function. "Information," he felt, was confusing because it already had too many connotations in English. So what should he name it? As Shannon told one of his colleagues at Bell Labs:

> I thought of calling it "information," but the word was overly used, so I decided to call it "uncertainty." When I discussed it with John von Neumann, he had a better idea. Von Neumann told me, "You should call it entropy, for two reasons. In the first place your uncertainty function has been used in statistical mechanics under that name, so it already has a name. In the second place, and more important, no one knows what entropy really is, so in a debate you will always have the advantage."[6]

Indeed, the terms *entropy* and *information* are terribly confusing and seemingly unrelated. How can information, the answer to a question, be tied to entropy, the measure of the improbability of the arrangement of stuff in a container? As it turns out, the two are much more tightly bound than even Shannon suspected in 1948. Information is intimately related to entropy and energy—the stuff of thermodynamics. In a sense, thermodynamics is just a special case of information theory.

The function Shannon derived was, roughly speaking, a measure of how unpredictable a string of bits is. The less predictable it is, the less able you are to generate the entire message from a smaller string of bits—in other words, the less redundant it is. The less redundancy a message has, the more information it can contain, so by measuring this unpredictability, Shannon hoped to be able to get at the information stored in a message.

6. Tribus and McIrvine, "Energy and Information," 180.

What's the most unpredictable string of **0**s and **1**s possible? In my pocket, I've got a great unpredictable-bit-stream generator—a coin. The flip of a coin is a completely random event, and *random* simply means "unpredictable." You've only got a 50 percent chance of guessing what any given coin toss will be. Moreover, you can't come up with rules that find a pattern in the string of coin tosses because there *is* no pattern. Here's a random string of sixteen bits; I just flipped a coin sixteen times and wrote down a **0** for heads and a **1** for tails: **1011000100001001**. This is a random, patternless stream of bits. There are no underlying rules that tell you what any given flip will be—or that will even give you a better than 50 percent chance of guessing what any given digit is. This is incompressible, and a random-looking stream like this therefore tends to carry sixteen bits of information; each symbol in the stream tends to carry one bit of information.

At the other extreme, imagine that my coin is biased: it always, 100 percent of the time, comes up tails. If I were to generate a stream of sixteen bits with it, it would look like **1111111111111111**. This is easily predictable; you have a 100 percent chance of guessing what any given flip of the coin or digit in the stream is. It is entirely redundant, so it doesn't carry any information; each symbol in this stream carries no bits of information.

What about something in between? What if my coin is weighted so that it comes up tails 75 percent of the time and heads 25 percent of the time? Sixteen flips of such a coin might yield something like **0101011111111111**. This is not *entirely* predictable, but since the coin is biased, if someone asked you to guess what any given digit was, you would be right roughly 75 percent of the time if you always guessed **1**. There is thus an underlying rule that helps you guess the outcome of any given flip; therefore, a stream like this is somewhat redundant, but not completely so—it can carry some information, but probably not a full bit per digit.

The more random—the less predictable—a stream of symbols is, the less redundant it is, and the more information it tends to carry per symbol. This is a seemingly paradoxical statement. How can something that is inherently random carry a message? Isn't randomness the *opposite* of purposeful information? Yes. But Shannon's point is that streams that *look* random—the ones that are the least predictable— are the ones that are likely to carry the most information per symbol. The nonrandom-looking ones, the predictable streams, are redundant and therefore probably carry less information per symbol than the random-looking ones do.

The reason I put weasel words like "tends to" and "probably" in the above analyses of the information content of a stream of digits is that I have been oversimplifying a little. This is a subtle point, but it is important. Shannon's analysis really is performed on the *source* of the message—a computer sending out electronic signals or a cell phone sending out voice data—rather than on an *individual* message itself. A source of data, such as an idling computer, that uses the rule "All digits you produce are **1**s" to generate messages will always yield the message "**11111111** . . . ." Every message from this source looks the same and will contain no information whatsoever. But a source of data that has no rules—in which **0**s and **1**s are equally likely and independent of each other—is inclined to produce "random-seeming" strings like "**10110001** . . . ." Unlike the "always **1**" source, which always produces the same message with no bits of information per digit, this "random-seeming" source can produce many, many different kinds of messages, each of which has one bit of information per digit. But—and here's the tricky part—the "random-seeming" source may well produce the message "**11111111** . . ."; it's very, very improbable, but it's possible.[7]

7. How can one message of "**11111111** . . ." contain no information and another message of "**11111111** . . ." contain lots of it? If the streams of digits are infinitely long, then there is absolutely *no* chance of a "random-seeming" source producing a message of all **1**s, thanks to a mathematical law known as the law of large numbers. So in the infinitely long message case, you can *always* distinguish a "random-seeming" source from an "always **1**" source by looking at a single message. In other words, there is no difference between the entropy/information content of a message and the entropy/information content of a message source. In the real world, though, messages are finite. There is a

That caveat aside, it makes perfect sense to talk about the information content in a stream of digits, but if you are going to do so—if you are going to measure the possible information stored in a set of symbols—you have to come up with some gauge of the predictability, the "random-seemingness" of that stream. Shannon came up with one. If p is the probability of a **1** in a stream of **0**s and **1**s, then the randomness is related to log p. Log p should seem familiar—it featured prominently in our analysis of the entropy of a container full of gas, and it's no coincidence that Shannon's measure of randomness is *precisely* the same function as Boltzmann entropy.[8]

Remember that we derived Boltzmann entropy by tossing marbles in a box. We then counted whether the marble wound up on the left side or the right side. This is equivalent to a coin flip; each marble landing in the box can land on the left or the right, just as a coin can land as a heads or a tails. Boltzmann entropy is a measure of the probability of each outcome in the marble-tossing experiment. The most probable ones, where about half the marbles land on the left and half land on the right, have the highest entropy; the least probable ones, where 100 percent of the marbles land on the left or 100 percent land on the right, have the lowest entropy. And the ones between these extremes, where, say, 75 percent of the marbles land on the left and 25 percent land on the right, have an intermediate entropy.

---

small probability that a "random-seeming" high-information source will produce a message that looks nonrandom. It can even look like one produced by an "always **1**" no-information source. This probability is *extremely* small—in an eight-bit message, the chance is less than 0.1 percent; in a sixteen-bit message, less than 0.0016 percent. In fact, it's just like tossing marbles in boxes. The probability of getting a message from a "random-seeming" source that looks like it comes from a nonrandom source is similar to the probability of having all or almost all of the marbles land on one side of the box. It's a possibility, but in reasonably large systems it's so improbable that it can be ignored. Therefore, in *most* cases—especially those where messages are sufficiently large or where a stream represents a sufficiently large collection of messages—the entropy/information content of a stream of digits is precisely the same as the entropy/information capacity of the source of that message. The equivalence is statistical, just as the second law of thermodynamics is statistical.

8. For a full examination of the different entropy functions, and a deeper explanation of the relationship between entropy and information, see appendix B.

This is exactly the same thing that we have seen in streams of digits. The case where 50 percent are 1s and 50 percent are 0s seem the most random, can carry the most information, and have the highest Shannon entropy. The case where 100 percent of the digits are 1s seems the least random, can carry the least information, and has the lowest Shannon entropy. An intermediate case, where 75 percent are 1s and 25 percent are 0s, is somewhat random, carries some information, and has a moderate Shannon entropy. (In fact, such a stream can carry about 0.8 bits per symbol.) Entropy and information are twins.

When Shannon realized that the entropy of a stream of symbols was related to the amount of information the stream tends to carry, he suddenly had a tool to quantify the information and redundancy in a message, which is, after all, what he set out to determine. He was able to prove, mathematically, how much information can be transmitted in any medium, via semaphore flags or smoke signals, by lamps in a belfry or by telegraph. Or how much information can be carried upon a copper telephone line. This is a surprising result: there is a fundamental limit to how much information you can transmit with a given piece of equipment. He also figured out how to deal with noisy connections between a sender and a receiver (noisy "channels"), and with methods of transmission that were not made up of discrete symbols but of continuous ones. His work led to the error-correcting codes that allow computers to operate. Shannon also figured out how much energy was required to transmit a bit from place to place under certain conditions.

Shannon's work opened up an entirely new field of scientific knowledge: the theory of communications and of information. For years, cryptographers had been trying to hide information and reduce redundancy without knowing how to measure them; engineers had been trying to design efficient ways of transmitting messages without knowing the limits that Nature puts upon their efficiency. Shannon's information theory revolutionized cryptography, signals engineering, computer science, and a number of other fields. But if that were all that information theory did, it would scarcely be a revolution on the scale of relativity and quantum mechanics. What gives information theory its true power is its close tie to the physical world. Nature seems to talk in terms of information, and only through information theory could scientists understand the message it was sending.

Shannon himself didn't concentrate on the tie between the abstract world of information and the concrete world of thermodynamics. Besides his work on information theory, Shannon did a mathematical analysis of juggling and got interested in cybernetics, artificial intelligence, and teaching computers to play games. Based upon discussions with the artificial-intelligence guru Marvin Minsky, he actually built what he called the "ultimate machine," which perhaps represents what will happen when machines learn to think.[9]

But other scientists were consumed by a question. Was Shannon entropy truly related to thermodynamic entropy, or was the similarity cosmetic? Just because Shannon entropy—the measure of information—looks, mathematically, exactly the same as Boltzmann entropy—the measure of disorder—it doesn't necessarily mean that the two are *physically* related. Lots of equations look the same and have little to do with each other; mathematical coincidences abound in science. But in fact, Shannon entropy *is* a thermodynamic entropy as well as an information entropy. Information theory, the science of the manipulation and transmission of bits, is very closely tied to thermodynamics, the science of the manipulation and transfer of energy and entropy.

9. Arthur C. Clarke described Shannon's "ultimate machine": "Nothing could be simpler. It is merely a small wooden casket, the size and shape of a cigar box, with a single switch on one face. When you throw the switch, there is an angry, purposeful buzzing. The lid slowly rises, and from beneath it emerges a hand. The hand reaches down, turns the switch off and retreats into the box. With the finality of a closing coffin, the lid snaps shut, the buzzing ceases and peace reigns once more. The psychological effect, if you do not know what to expect, is devastating. There is something unspeakably sinister about a machine that does nothing—absolutely nothing—except switch itself off." Quoted in Sloane and Wyner, "Biography of Claude Elwood Shannon."

In fact, information theory banished the most enduring paradox in thermodynamics—Maxwell's demon—once and for all.

Maxwell's demon was such trouble because it seemed to punch a hole in the second law of thermodynamics. The tiny, intelligent demon-being—whether it be man or machine—seemed as if it could exploit the random, statistical element in matter to decrease entropy without any expenditure of energy. If this were true, even in principle, the second law of thermodynamics had a loophole. As soon as someone figured out how to manufacture a demon, the world would be supplied with endless amounts of energy, and the entropy of the universe wouldn't change at all. Of course, the demon had to be stopped.

The first step to dispelling the demon came before Shannon formalized information theory, but it was related to information nonetheless. In 1929, the Hungarian-born physicist Leo Szilard analyzed a modified version of Maxwell's demon—instead of opening or closing a shutter, the demon simply had to decide which side of a partition an atom is on—but the physics underlying Szilard's demon was precisely the same as Maxwell's. Through his detailed analysis, Szilard realized that the act of measuring the position of the atom (or in the Maxwell case, the speed of an incoming atom) must, in some way, increase the entropy of the universe, counteracting the demon's reduction of the universe's entropy. When a demon performs a measurement, he is getting an answer to a question: Is the atom on the right side of the box or the left side of the box? Is the atom hot or cold? Should I open a shutter or not? So a measurement is an extraction of information from the particle. That information does not come for free. Something about that information—either extracting it or processing it—would increase the entropy of the universe. In fact, Szilard calculated that the "cost" of that information was a certain amount of useful energy—more precisely, $kT \log 2$ joules for every bit of information, where T is the temperature of the room that the demon is in and k is the same constant that Boltzmann used in his entropy equation. Using that useful energy increases the entropy of the box. So the process of obtaining and acting upon that information increases the universe's entropy, counteracting the demon's efforts to decrease the entropy of the box by $kT \log 2$ joules for every bit of information it obtains and acts upon.

In 1951, another physicist, Léon Brillouin, took the next step. Inspired by Shannon's theorems, he tried to figure out, more specifically, what the demon was doing that increased the entropy of the box. Brillouin realized that a major obstacle was that the demon was blind. The box is dark, and the demon wouldn't be able to see the atoms, so Brillouin gave the demon a flashlight to help illuminate the skittering particles. The demon would shine the flashlight at an incoming particle, and when the beam reflected off the atom, the demon would act upon the information it received and decide whether to open or close the shutter. Brillouin calculated that the act of bouncing light off an atom, detecting the reflected light, and acting upon that information would increase the entropy of the box at least as much as the demon could decrease it. More important, because the process of extracting and acting upon Shannon-like information to answer a question about an incoming atom—Is it hot or cold?—increased the thermodynamic entropy of the box, Brillouin concluded that thermodynamic entropy and Shannon entropy were directly related. You could use the language of information theory instead of the language of thermodynamics to analyze the behavior of a box full of gas.

The laws of information theory provide a slightly different perspective from that of the laws of thermodynamics. For example, take a box full of gas. In thermodynamics language, we can apply energy (by running that same air conditioner or employing Maxwell's demon) to separate the hot molecules from the cold molecules, reducing the box's entropy and making one side of the box hot and the other side cold. After we stop applying energy, though, the box quickly returns to equilibrium. The entropy of the box increases again until the system is in equilibrium.

Using the language of information instead of that of thermodynamics, the exchange seems a little different. At the beginning, the box is in a state of equilibrium. We can apply energy (again, by running the air conditioner or employing Maxwell's demon) to gather and process information about the molecules in the box. That processing changes the information stored within that box. Maxwell's demon, according to Brillouin, was transferring information to the container, separating hot from cold molecules.[10] Once you stop applying energy, though, that stored information leaks out into the environment—for Nature, it seems, attempts to dissipate stored information just as it attempts to increase entropy; the two ideas are exactly the same.

It seems obvious, but not everyone agreed at the time. A number of scientists and philosophers of science objected to Brillouin's argument and the link between information and thermodynamic entropy. They argued that the similarity between the two entropy formulas was coincidental and that the two weren't related, and those objections continue to this day. And in fact, with clever measurement schemes, you can detect an atom with arbitrarily little entropy production and energy consumption. However, a powerful argument forged an even tighter link between thermodynamics and information theory, and put an end to Maxwell's demon.

The insight came from an unexpected quarter: computer science. Recall that in the 1930s, Alan Turing, the soon-to-be Enigma codebreaker, proved that a simple machine that could make a mark on a tape, erase a mark on a tape, and move the tape around was able to do anything that any computer could conceivably do.[11] If you think of a mark on a tape as a **1** and an erased portion of a tape as a **0**, you can recast Turing's proof in another way: you can do anything any computer can do by storing, manipulating, and erasing bits. Since Shannon proved that bits are the fundamental units of information,

10. Brillouin's notation—that the higher a system's entropy, the lower the information it contains—seems to be the reverse of what I implied in the marbles-in-a-box derivation. In truth, the two are the same; see appendix B for an explanation.
11. As we shall see in a later chapter, we're talking about "classical" computers here, not quantum computers.

processing information was nothing more than manipulating bits, something that the Turing machine was designed to do. Conversely, a computer was nothing more than an information-processing machine, and in processing information it became subject to the laws that Shannon set out. Manipulating, processing, and transmitting information was linked to the consumption and production of energy and entropy; manipulating energy and entropy was the essential function of an information-processing machine, such as a Turing machine, a computer, or a brain. The ideas were intimately linked—understand the relationship among entropy, energy, and information and you might begin to understand how computers and humans think. So, in the wake of Shannon's discoveries, scientists set out to determine how much energy and entropy a computer consumed or produced when carrying out its manipulation, as a first step to understanding how computers and brains worked.

In 1961, the physicist Rolf Landauer came up with a surprising answer to how a computer (or a brain) uses energy to do its information processing (or thinking). It turns out that you can add bits without consuming energy or increasing the energy of the universe. You can multiply bits. You can negate them. But one action in a computer generates heat, which then dissipates into the environment, increasing the entropy of the universe. That action is erasing a bit. Erasure is the action in a computer's memory that costs energy.

Landauer's principle, as it came to be known, is rather counterintuitive, but it comes from solid physical principles. Instead of a silicon chip, let's use a two-meter-long pool table as our computer's memory. A half-kilogram billiard ball will be our bit; if it's at the left cushion of the table, the ball represents a **0**; if it's at the right, it's a **1**. We can do a simple operation on this bit in memory. The only rule is that there can be only a single recipe for an operation, and this single recipe must work no matter whether the ball is on the left or on the right. The instruction set must be symmetric: we can't give a **0** ball a different recipe to follow than we give to a **1** ball.

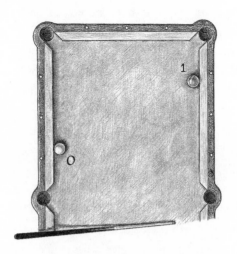

Storing a 1 and a 0 on a pool table

As an example, let's take the operation "negate": if the memory has a **0** in it, change it to a **1**; if it has a **1**, then change it to a **0**. This is a pretty easy thing to do with our pool table memory. Here's the recipe: give the billiard ball one joule of energy so that it moves to the right at two meters per second. One second later, stop the ball and reclaim that one joule of energy. This is a single set of instructions, and it works for both of our billiard balls.

If our memory starts out as a **0**, the pool ball at the left cushion moves to the right at two meters a second. Exactly one second later, it hits the cushion—and at that very moment we stop the ball, taking away its energy. The **0** has become a **1**. On the other hand, if our memory starts out as a **1**, the ball starts off at the right cushion. It starts moving right at two meters a second but is immediately reflected off the cushion and moves left at two meters a second. Precisely one second later, when we remove its energy, it has moved across the table and is touching the left cushion. The **1** has become a **0**. Ideally, with a perfect table, no energy is lost. In both cases, we reclaim the joule that we have put in; we have negated our memory without consuming or dissipating energy.

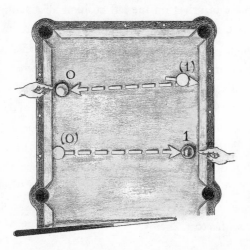

---

*Negating a **1** and a **0** on the pool table (reclaiming the energy by hand)*

Now let's come up with a recipe for erasing our pool table memory. No matter whether we start off with a **1** or a **0** stored in memory, we want to end up with a **0**, the billiard ball sitting at the left cushion. This is not so easy. We can't use the negate trick we used before; it works if we start off with a **1** in memory, but it fails if we start off with a **0**. And since we're only allowed to write a single set of instructions that applies for both balls, we can't say, "negate if the ball is on the right but do nothing if the ball is on the left"—that would be giving a different instruction to each ball.

But there is a way to do it with a single instruction. We have to modify the pool table slightly. Let's put a piece of energy-absorbing plush velvet on the left cushion; when a ball smacks into it, the velvet absorbs all the energy and brings the ball to a halt. Now, let's do the negate trick as before, but leave off the last instruction to reabsorb the energy one second later; all we do is give the ball a knock, putting one joule of energy into it and making it roll to the right.

If the ball starts off at the right cushion, if the memory is a **1**, it will hit the right cushion immediately and roll to the left. One second later, it hits the piece of velvet on the left cushion, dissipating the ball's energy and bringing it to a halt at the left cushion. After two seconds, the ball hasn't moved; it is still a **0**. Our recipe has turned a **1** into a **0**, so we know it works when we start off with a **1**. But what about when we start off with a **0**, with a ball at the left cushion? Well, a ball on the left will immediately start rolling to the right because of the energy we put in. One second later, it hits the right cushion, bounces off, and starts rolling to the left again. One second later, it hits the piece of velvet. The energy dissipates, and the ball comes to a halt at the left cushion. The **0** flips back and forth but winds up as a **0** two seconds later—and stays that way. Our recipe works for a **0** as well as for a **1**. But it comes with a cost. Energy.

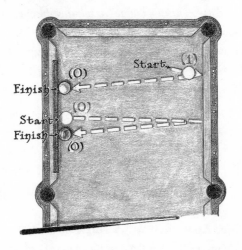

Erasing a 1 and a 0 on the pool table

With the negate command, we put in a joule of energy at the beginning of the recipe and we retrieved a joule of energy at the end; no energy was spent in changing a **0** to a **1** and vice versa. (The negate recipe even works with the modified, velveted pool table. We can retrieve the energy at the very moment a ball begins to touch the velvet, before any energy is lost.) But with the "erase" command, setting everything to **0**, we have to let the velvet bring the balls to a halt. The velvet acts as a brake; it takes the joule of energy from the ball— whether it started off on the left or on the right—and dissipates that joule of energy into the environment in the form of heat. That's what brakes do. We have no choice but to use a mechanism like this; we can't put a "retrieve energy" into our instruction set for erasure, because that energy retrieval makes us unable to have both balls wind up at the left cushion at the end of the recipe. Only by adding the strip of velvet, only by giving up on retrieving the energy we put in, can we execute an erase command that is valid when our memory starts out with either a **0** or a **1**. Erasing memory causes heat to flow into the environment. This is Landauer's principle.

The act of erasing a bit in memory releases heat, which dissipates into the environment. As soon as that energy dissipates, it increases the universe's entropy just as surely as if a small gobbet of helium were to dissipate throughout a container. The processing of information is a thermodynamic process—and vice versa. Deeper still, the crux of Landauer's principle, the idea that erasure increases the entropy of the universe, is that erasure is an *irreversible* operation. If you take a bit in memory and erase it, letting the heat dissipate away, there is no way to recover that bit. This is different from an operation like negation, which can easily be reversed by a second negation, or like addition, which can be reversed by subtraction. Reversible operations don't increase the entropy of the universe; irreversible ones do. The entropic arrow of time applies to the manipulation of bits just as it does to the motion of atoms. You are unable to reverse a film of an irreversible process—informational or physical—when the entropy of the universe has changed.

In 1982, the IBM physicist Charles Bennett took the final step that would dispel Maxwell's demon forever. If you put a demon inside a box and gave it instructions to make one side of the box hot and the other cold, the demon would be making decisions about whether or not to open or close a shutter; it would be making binary decisions that help it achieve the goal of reversing entropy in the box. The demon would be, at heart, an information-processing machine—a computer—programmed with the instructions you gave it. And since a Turing machine can do whatever any computer can do, you can have a Turing machine act as the demon. The Turing machine will have to measure the speed of an atom somehow, write a bit on tape that records the result of that measurement, and then execute a program that uses that bit in memory to decide whether or not to open or close the shutter. But the act of writing that bit implicitly requires that you erase the memory position where you are writing to get rid of the data from the previous measurement. Even if you have lots of memory available—and you can switch to a fresh, unused section of memory for each new atom—you will run out of memory sometime unless you've got an infinite amount of it. Since there are a finite number of particles in the universe, you can't have infinite memory; the demon will run out of tape sometime and will eventually have to erase memory to clear up room for new measurements. For a while, the demon can run, filling up its memory with information, but as soon as it runs out of tape, it produces more entropy by liberating heat into the universe than it removes by separating hot atoms from cold atoms in its box. Bennett proved that the demon always has to reduce entropy in the container at a cost: a cost of memory, and then at the cost of raising the entropy of the universe. There was no free ride, no perpetual motion machine. Maxwell's demon was dead at the age of 111.

The greatest thermodynamic paradox was, in truth, a paradox about the manipulation of bits of information. Shannon didn't set out to solve the paradox of Maxwell's demon or to figure out the power consumption of a Turing machine, but the connections among thermodynamics, computers, and information were much stronger than Shannon imagined when he founded the discipline of information theory.

It was much deeper than even Brillouin, who vociferously argued that the Shannon and Boltzmann entropies were closely related, could know. As Landauer wrote in 1996,

> Information is not a disembodied abstract entity; it is always tied to a physical representation. It is represented by an engraving on a stone tablet, a spin, a charge, a hole in a punched card, a mark on paper, or some other equivalent. This ties the handling of information to all the possibilities and restrictions of our real physical world, its laws of physics and its storehouse of available parts.[12]

The laws of information had already solved the paradoxes of thermodynamics; in fact, information theory consumed thermodynamics. The problems in thermodynamics can be solved by recognizing that thermodynamics is, in truth, a special case of information theory. Now that we see that information is physical, by studying the laws of information we can figure out the laws of the universe. And just as all matter and energy is subject to the laws of thermodynamics, all matter and energy is subject to the laws of information. Including us.

Though living beings seem as if they are inherently different from computers and boxes of gas, the laws of information theory still apply. We human beings store information in our brains and our genes just as a computers store information in their hard drives, and in fact, it seems that the act of living can be seen as the act of replicating and preserving information despite Nature's attempts to dissipate and destroy it. Information theory is revealing the answer to the age-old question, What is life? That answer is quite disturbing.

12. Quoted in Leff and Rex, eds., *Maxwell's Demon 2*, 335.

CHAPTER 4

LIFE

Instead of asking which came first, the chicken or the egg, it suddenly
seemed that a chicken was an egg's idea for getting more eggs.

—Marshall McLuhan,
Understanding Media

In 1943, in the midst of World War II, an eminent physicist, Erwin
Schrödinger, presented a series of lectures at Trinity College in
Dublin. Schrödinger was famous for deriving the fundamental laws of
the quantum realm. You've probably heard of Schrödinger's cat, which
is a seeming paradox based upon the difference between the quantum
laws of the subatomic realm and the classical laws of the everyday world. However, the subject of Schrödinger's lectures wasn't the
weirdness of quantum mechanics, nor was it the behavior of nuclear
matter, a topic already of great interest to scientists at Los Alamos,
New Mexico. Schrödinger, the physicist, lectured upon a subject that
seemed far removed from the quantum physics that made his name.
He talked about answering the fundamental question of biology:
What is life?

What makes a rat or a bacterium different from a rock or a drop of
water? Despite millennia of trying, philosophers and scientists failed,
over and over again, to come up with a satisfying answer. In his lectures, Schrödinger tried to tackle the question because he saw a deep

connection between the seemingly unrelated fields of quantum theory and the philosophy of the nature of life. The terminology hadn't been invented yet—Shannon's theory was half a decade away—but Schrödinger sensed that this connection had to do with what would be known as information.

Looking at it from a physicist's perspective, a living organism, Schrödinger noticed, is continuously fighting off decay. It maintains its internal order despite a universe that is always increasing in entropy. By eating food, by consuming energy that ultimately comes from the sun, the organism is able to keep itself far from equilibrium: from death. And though Schrödinger didn't use the phrasing that information theorists use—he was speaking before information theory was born, after all—he explained that life was a delicate dance of energy, entropy, and information. He, like all other scientists of the day, didn't know what this information was or where it resided, but he sensed that the essential function of living beings is the consumption, processing, preservation, and duplication of information.

This information of life is much more than what is responsible for consciousness and the information that is being crunched in our brains. Information is responsible for all life on Earth. The laws of information guide every living creature, down to the lowliest bacteria and to the smallest living particles in the world. Every cell in our bodies is packed with information. We eat so that we can process that information. And our whole being is co-opted to transmit information from generation to generation. We are slaves to the information inside us.

If we are to understand what life is and how it came to be, we must understand what that information is telling us. Shannon's theory tells us how to measure and manipulate that information—and what laws that information must obey when stored within a living body. Shannon's information theory made the question of life a question for physicists as much as it is for biologists, philosophers, and theologians.

When Schrödinger gave his lectures in Dublin in 1943, scientists didn't know all that much about the genetic code. It was a full decade before James Watson and Francis Crick discovered the double-helix structure of DNA. Biologists knew that traits were passed down from generation to generation. They knew that traits were somehow encoded in units known as *genes,* and that *something* in the cells, some sort of molecule, was somehow responsible for these genes. Biologists and physicists knew roughly where these molecules were, and roughly how big they were.

Most scientists of the day—Schrödinger included—thought that proteins were the molecules in question, the carriers of genetic information. They were wrong. Biologists and physicists now know that DNA, deoxyribonucleic acid, is the once-mysterious molecule that carries the genetic code. It is a molecule whose purpose is to store information, protect it from dissipation, and duplicate it when necessary. But Schrödinger's talks were about the message, not the medium, and about that he was correct. And though Schrödinger was forced to talk about the genetic code in terms of the dots and dashes of Morse code, we can talk of it in terms of information.

Even though Schrödinger was confused about what molecule stores the information in our cells, and even though he didn't have information-theoretic language with which to couch his talk, the core of his message—bafflement—is still valid. For Schrödinger was having difficulty understanding the amazing permanence and resilience of the information stored in our cells. Even though it is duplicated over and over, passed down from generation to generation, this information changes very little over time. The information is preserved, kept safe from dissipation.

This is not the way Nature usually behaves. Entropy naturally increases in a system that is left to its own devices. A box of gas quickly settles into equilibrium. Information tends to dissipate; stored information eventually diffuses throughout the universe. Information spreads out, especially in big, complex, warm systems like living creatures. And once a creature dies, it immediately begins to decay; its flesh falls apart and so do the molecules that make up that flesh. With them, the creature's genetic code is, over time, scattered to the winds. Somehow, being alive allows living beings to preserve their information, seemingly flouting entropy for a short time. Once a creature dies, though, that ability is lost forever, and entropy wins as the creature's information is scattered.

Scientists now know a lot more than Schrödinger did. In 1953, Watson and Crick showed that our genetic code is inscribed upon long, stringy, double-stranded DNA molecules. The most important part of the molecule, information-wise, is where the two strands join together in the middle. It is there that each strand contains its message. This message is not written in binary code; it is not a code of **0**s and **1**s or **T**s and **F**s. It is a quaternary code, a code that has four symbols. Each symbol is one of four chemicals, or bases: adenine, thymine, cytosine, and guanine. If you were a molecular-size creature and you rappelled down one strand of DNA, you would see a sequence of these chemicals in a well-defined order, say, **ATGGCGGAG**. Attached to this strand, base touching base, you would see another strand equal and opposite to the first.

Adenine and thymine are chemicals that complement and bind to each other; cytosine and guanine are also complementary and bind together. The other strand, which, in fact, runs in the opposite direction to the first, replaces each chemical in the first strand with its complement. So, in our example, the complimentary, *antisense* strand has the sequence **TACCGCCTC**, which binds nicely to the sequence **ATGGCGGAG** like so:

Since the two strands can be separated from each other, the DNA molecule effectively has two copies of the same information. Information is the proper word to use, because, in fact, DNA is storing information in the Shannon sense. Shannon's theory applies to any string of symbols, and like any other symbols, DNA's quaternary code can be reduced to a string of bits, of 0s and 1s—two bits for each chemical. (For example, we can represent A, T, C, and G by 00, 11, 01, and 10 respectively.) As important as it is for life, from an information theorist's point of view DNA is no different from any other medium that can store information. If you can figure out how to manipulate the information on a strand of DNA, you can use it as the "tape" in a Turing machine; if you have the ability to read and write to DNA strands, you can turn it into a computer. In fact, this has been done many times.

For example, in 2000, Laura Landweber, a biologist at Princeton, created a "DNA computer" that solved a famous computer-science puzzle known as the knight problem. Given a chessboard of a certain size—in Landweber's case three squares by three squares—what are all the possible ways you can put chess knights (which move in an L-shaped pattern) on the board and yet have them unable to attack each other?

Landweber exploited a number of tools that biologists had created over the years to manipulate DNA and a related information-containing molecule known as RNA (ribonucleic acid). Scientists had developed procedures—using enzymes and chemicals—for reading the code inscribed on DNA molecules, for writing any desired set of symbols on a given strand of DNA, and for duplicating that information manyfold. They also had the ability to cleave and destroy molecules that contained an unwanted sequence of symbols. These are all operations that manipulate information. In fact, these operations are enough to create a primitive computer out of DNA.

Landweber did it with a "brute force" approach. First, she synthesized eighteen different stretches of DNA, each consisting of fifteen base pairs. Each stretch represented a bit for a particular space—a "knight" or a "blank," a **1** or a **0**, for each of the nine positions on the board. (For instance, **CTCTTACTCAATTCT** meant that the upper-left corner is blank.) She then created a "library" of millions of DNA strands representing all possible configurations of the board—that is, every possible permutation of knights and blanks. Landweber then methodically eliminated the permutations in which one knight could capture another, mincing all the non-solution-bearing molecules with cleavage enzymes.[1]

The procedure was equivalent to a series of logical operations on a computer. After taking clumps of DNA goop, writing information onto them, and manipulating the information on the molecules in such a way that the DNA executed a logical program, Landweber had a beakerful of DNA strands that held solutions to the knight problem, just as surely as a computer that had executed the logic would have held an answer in its memory bank. When she read out and deciphered the code of forty-three of those strands—the equivalent of asking the computer to print the contents of its memory bank—she discovered that forty-two held valid solutions to the problem. (One had an incorrect solution: a mutation.) Landweber had executed a computer algorithm on a strand of DNA.

By Nature's standards, though, Landweber's methods were very crude

1. The enzymatic algorithm was easy to perform because the knight problem can be reduced to a set of simple logical statements. One statement might be: "*Either* the upper-left corner is blank, *or* the two squares that a knight threatens from that position must be blank." To satisfy that statement, Landweber split the library into two. Into one jug she poured an enzyme that targeted the sequence that meant "there is a knight in the upper-left corner." To the other jug she added two enzymes that targeted the sequence that signaled the presence of a knight in the two threatened positions. After the broken fragments were all weeded out, neither jug contained a strand that included sequences that had *both* a knight in the upper-left corner and a knight in one of the two squares threatened from that position. And then Landweber combined the jugs; there was no sequence in the library that had a knight in the upper-left corner *and* a knight in one of the positions that the knight threatened. She then repeated the procedure for all the squares—either there's no knight in square 1 or there's no knight in squares 6 and 8; either there's no knight in square 2 or there's no knight in squares 7 and 9; and so forth. After all the splitting, cleaving, and combining, no strands were left where one knight threatened another.

and clumsy; her toolbox only had a small number of ways she could manipulate the information on DNA. She could force information-containing molecules to reproduce; she could cleave them in two and destroy them; she could write a particular code onto a strand of DNA from scratch. But she couldn't do other basic functions that a Turing machine should be able to do. For example, while she could build a code from scratch, once it was written she couldn't edit it—she couldn't pull out, say, a **C** from a strand and replace it with an **A**. She could not correct informational errors—mutations—that occurred during processing.

Nature has tools to do all of these things. Enzymes—proteins in the cell—continually monitor DNA molecules, looking for mutations and editing them out. Each cell in our body is home to thousands of these proteins, which manipulate the information in our DNA—duplicating it, writing to it, reading from it, editing it, transferring it to other media, and executing instructions written upon it. And the instructions for making and regulating these proteins is encoded in DNA as well. In a sense, at the heart of each of our cells is a computer that runs off of the instructions contained on the DNA molecule. Yet, if a computer is ticking away in each of our cells, running on the program stored in our DNA, what does that program do?

There's a huge effort to decipher genetic code of all sorts of organisms—to read out the details of these computer programs. But even without knowing the exact details of all of these programs, many evolutionary biologists already suspect that all the programs are doing precisely the same thing. They are executing one simple command.

Reproduce. Duplicate your information.

Sure, the programs go about this task in very different ways, but the goal is always the same. Reproduction. All else is decoration—decoration that helps the program achieve its ultimate goal. Bodies—and their arms and legs and heads and brains and eyes and fangs and wings and leaves and cilia—are just packaging for the information contained in an organism's genes, packaging that makes it more likely that the information contained in the packaging will get a chance to replicate itself.

This is an incredibly reductivist way of viewing living creatures. It is probably different from what you learned in biology class, where evolution is portrayed as *individuals* trying to reproduce—where the fittest *organisms* survive, and the genes' function is to make their organisms fitter. Not all scientists view genetics in quite this way, but many biologists argue that an organism's genes, the information in its cells, are not "trying" to make a fitter organism: they are simply trying to duplicate themselves.

It's a subtle point. It is not the individual that is driving reproduction; it is the information in the individual. The information in an organism has a goal of replicating itself. While the organism's body is a by-product, a tool for attaining that goal, it is just the vehicle for carrying the information around, sheltering it, and helping the information reproduce. That the *organism* reproduces is just a by-product of the information's duplicating itself . . . sometimes.

An organism's information can sometimes reproduce itself *without* having its vehicle organism reproduce. Consider ants, for example. In a typical ant colony, only one organism is fertile—the queen. Only she is reproducing; only she is laying eggs. All the other thousands upon thousands of ants in the colony are (more or less) sterile and unable to reproduce. Yet these sterile ants tend the queen's eggs and rear them to adulthood. Even though they are not the parents of the eggs, they care for the queen's brood.

Almost none of the organisms in this colony will ever produce young. They give up their reproductive ability and are completely subjugated to rear another individual's young. Yet the information inscribed on their genes instructs them to obey the queen and forfeit their hopes of reproduction. If, in fact, the *individual* is in control, if the individual is what's trying to reproduce, this strategy makes no sense. But if the *information* in the organism is in control and is the entity trying to reproduce, the sterile ant's behavior begins to look rational.

If you are a worker ant in the colony, your mother is the queen, and your mother's genes contain almost all of your genetic material—including the "obey the queen" gene.[2] All of her brood—your sisters—also have the "obey the queen" gene in their DNA. So, by following the instructions in the program, by obeying the queen and tending the brood, a sterile worker ant is helping the "obey the queen" gene flourish. From the individual's point of view, the individual has failed to reproduce, but from the "obey the queen" gene's point of view, the gene has succeeded: it gets to reproduce itself, even though most of the individuals that carry it don't. So, sterility makes perfect sense for the information in the ant's genes, even if it doesn't make sense for each individual ant.

This is an example of how effects that the genes have on their vehicle organisms aren't "intended" to make the organism fitter. A sterile ant is less fit, in a Darwinian sense, than one that isn't. However, genes often *do* have that effect. Genes for poison and fangs probably help the rattlesnake pass on the genes for poison and fangs; by having a beneficial effect on their hosts, these genes increase the likelihood that the host organism—and the information it contains—reproduces. But not all genes have a beneficial effect on their host organism. Some genes are downright harmful—more harmful even than sterility—yet they, like other genes, are trying to replicate themselves.

There is a gene that sometimes shows up in mice known as the *t* gene. The *t* gene has no apparent beneficial effect; in fact, it is often fatal. If a mouse happens to have two copies of *t* in its genetic programming, the mouse dies or is unable to reproduce. But if a mouse has a single copy of *t*, nothing happens—well, not quite nothing.

2. The "obey the queen" gene is a convenient fiction. With many traits and behaviors, such as "obey the queen," no single, individual gene can be pinpointed as the cause. They are the products of complex interactions of instructions in the genetic code together with cues from the environment. Nevertheless, the overall argument I'm making remains the same, whether the program is a simple, single gene or something considerably more complex. Thus, I will refer to things like an "obey the queen" gene, even though the behaviors and traits I talk about are seldom controlled by something so simple as a single gene.

The *t* gene has a peculiar property: it is *really* good at getting itself replicated. Somehow, during the cell divisions that lead to the production of sperm, the *t* gene pushes to the front of the line and gets itself into almost all of the mouse's sperm. Ordinary mouse genes tend to wind up in 50 percent of mouse sperm cells, but the *t* gene manages to make it into 95 percent of the sperm. The *t* gene is a chunk of information that is particularly good at reproducing itself, and it does it with abandon.

If a mutation creates a *t* gene in a male mouse, the *t* gene replicates itself over and over as the mouse and his progeny reproduce. It winds up in the mouse's children. And his children's children. And his children's children's children. The *t* gene quickly runs rampant through the mouse's family and then through the entire mouse population. But as the *t* gene executes its program over and over again, it begins to destroy the mouse population that carries its information. The gene rapidly becomes ubiquitous in a mouse population, so after a few generations two mouse parents are probably going to have the gene. This means that it is very likely that their offspring will have two copies of the *t* gene and die. According to the biologist Richard Dawkins, there is evidence that the *t* gene has even caused populations of mice to go extinct.[3]

All the *t* gene "cared about" was replicating itself, even though executing its "replicate yourself!" program was harmful to the organisms that carried that information. In the long run, the *t* gene wipes out the population of mice—and itself—but the gene is unable to stop running its program or to temper its inexorable drive to reproduce, reproduce, reproduce. The *t* gene is truly selfish; it replicates itself despite the great peril it causes for its host organism.

In a sense, the genes are constantly battling with one another, trying to get themselves reproduced. But this battle is a complex one;

3. Dawkins tells the story of the *t* gene (as well as a number of other reasons for believing that organisms should be considered the vehicles for the information inside them) in his famous book *The Selfish Gene*.

often collaboration has better results than competition. Many genes have adopted a "strategy" of cooperating with one another. Genes for fangs and poison tend to be associated with genes that allow the organism to digest another animal; you seldom see an herbivore armed with an offensive weapon like a poison bite. Though the information for fangs and the information for a carnivore's digestion are unaware of each other's presence, they each enhance the other's chance of reproduction if they are present together. Hence, the two genes "cooperate" with each other. (Of course, genes aren't conscious entities, so they can't really "cooperate" or "fight" or "intend." But since these programs do have a "goal" of sorts—reproduction—and various different means of attaining the program's goal—by giving a host organism poison fangs or by ensuring its transmission in sperm—anthropomorphizing a gene is a shorthand way of describing the types of interactions that different genes can have with each other when executing their programs.)

But not all genes cooperate. The *t* gene, for example, reduces the viability of the host organism, the mouse, reducing the chances for all the genes in the mouse to reproduce. Inside every organism there's a complex battle between genes as they each try to get themselves replicated, and from the gene's point of view an organism is just a vehicle that allows the gene to achieve its goal. Indeed, to the information inside us, the vehicles might be disposable; many genes eventually abandon their original vehicle for another, more convenient one. Many genes in modern-day creatures are merely hitchhikers that organisms have picked up along the way.

Nestled in one of our chromosomes—one of the twenty-three pairs of genetic-information packages in the nuclei of our cells—is a sequence of genetic code that was planted there by just such a hitchhiker. Sometime in the distant past, this hitchhiker infected us, forced its way into our cells, snipped apart our genetic code, and inserted its own instructions. In 1999, biologists discovered the traces of this ancient infection. It was a foreign code—the instruction set for an entire fossil virus—which forces our bodies to produce proteins that the virus desires, rather than what our cells themselves need.

In fact, the information in each of our cells is riddled with fossil, hitchhiking genes. Our bodies produce these *human endogenous retroviruses,* HERVs, because the code has been inserted in our genome, not because it has any beneficial effect on the organism itself. Millennia ago, the virus genes procured themselves a free ride; as humans reproduce, the virus genes reproduce as well. The human organism is merely a tool to this viral invader. We get no apparent benefit from the hitchhiker, and there is some evidence that it can do harm.

Luckily, some of these hitchhikers do have a beneficial effect; we owe our high-energy existence to one ancient hitchhiker. Each of our cells—indeed, each animal and plant cell—has a number of power plants inside it, little features known as mitochondria. We could not live without them. Mitochondria are responsible for extracting almost all the energy our cells need from chemicals and converting that energy into a usable form. There is good evidence that these mitochondria are actually bacterial hitchhikers that somehow injected themselves into our single-celled progenitor organisms billions of years ago. For one thing, mitochondria have a completely separate set of DNA from the stuff that is stored in the center of our cells; they are carrying a set of instructions entirely different from those in the cell nucleus.[4] Every cell in our bodies—skin cells, nerve cells, liver cells, kidney cells—is a schizophrenic, double creature because of the mitochondria inside. Every time a cell divides, it passes on mitochondrial DNA as well as its own DNA. Mitochondrial DNA is along for the ride.

The original creatures that gave us these snippets of information and injected them into our ancestors' cells—the virus responsible for HERV genes and the bacterium-like creature responsible for our

---

4. Because the mitochondria are hitchhikers, they can dispense with creating some of the important proteins that are responsible for cellular machinery. Human mitochondrial DNA contains about 33,000 bits of information, considerably less than what is contained in the string of letters that make up this chapter.

mitochondrial DNA—are extinct, so far as scientists know. Yet the information they carried is still with us. The information has jumped vehicles, and when the original organism died out, the information survived.

This leads to perhaps the most powerful argument that the information in our genes—not the organism that protects that information—is the fundamental element that's reproducing and surviving in the game of life. That argument is immortality. The information in our cells is essentially immortal, even though every single one of our cells, even those not yet born, will be dead in less than a hundred years. Much of the information in our genes is billions of years old, passed down from organisms that floated in the primordial ooze that covered the Earth when it was still young. Information not only can survive the death of the individual it resides in, it can also survive even the *extinction* of its host organism. This may be the answer to the eternal question, Why must we die? We don't. We are immortal. The catch is that the "we" in question is not our bodies or our minds; it is the bits of information that reside in our genes.

Though this line of argument seems to get closer to answering the question, What is life? it does not address Schrödinger's bafflement. Entropy degrades a device that stores classical information: computer hard drives get corrupted, books fade, and even stone etchings weather away. Nature tries to take information and spread it around the universe, rendering it inaccessible and useless. Yet the information in our genes is able to resist the ravages of time and entropy, the arrow of time. This is what so astounded Schrödinger and made him wonder about the nature of life. Immortality requires protection from entropy, yet the laws of thermodynamics say that entropy is inexorable. How, then, can life exist at all?

On a purely physical level, it is not too much of a puzzle. Just as a refrigerator can use its engine to reverse entropy—locally—by keeping its insides colder than the room it is in, the cell has biological engines that are used to reverse entropy—locally—by keeping the information in the cells intact.

There are thousands of enzymes in each cell that manipulate the information at the cell's core. There are duplicators, editors, and error checkers, performing the functions that you would expect a typical computer to be able to do. In fact, the double-helical structure of DNA is a particularly nice and stable storage medium for information because there are two copies of the information, one on each strand. Most errors can be caught by comparing the two strands; if there is a mismatch, then an error must have occurred. Perhaps a chemical **A** was accidentally swapped for a chemical **C**, or maybe one of these bases was erroneously doubled. The enzymes in our cells, the little molecular machines, continually scour the DNA strands in search of a mismatch or some other error. When they find one, they snip out the offending segment and replace it.[5]

Nature's random probings, such as bouncing errant molecules off the double helix or irradiating it with various types of photons, tend to cause the information on the DNA to dissipate. Such events will strip electrons and atoms from the DNA, cause kinks and bends and mismatches, and wreak other sorts of havoc. Nevertheless, the error-checking mechanisms in our cells are largely able to keep the information intact. At a cost. An energy cost.

Just as a refrigerator needs energy to stave off the effects of entropy—to keep part of a room cold and part of it hot—the molecular motors need, at some point, to consume energy if they are to oper-

---

5. These error checkers are very, very good, but they're not perfect. Once in a while, they fail to catch an error, which winds up being duplicated when the cell divides. This is a mutation. Often, mutations are harmful, causing an unwanted effect, perhaps even killing the organism that sports the mutation. In a sense, this is the final error-checking mechanism. Mutations to genes that are essential to the organism's survival are unlikely to be passed on (because they will probably mess up that essential function), but mutations to nonessential information (such as hitchhikers' stuff or extra copies of genes) don't have this last error check. This means that the nonessential information is less stable from generation to generation; it is more likely to contain a mutation. And in the rare event that a mutation has a *beneficial* effect, it becomes more likely to be passed on, because the host organism benefits from the gene's expression.

ate. For example, one enzyme, which detects a bulge in a DNA strand caused by two neighboring thymines linking to each other rather than to the adenines on their complementary strand, is activated by absorbing a photon of ultraviolet light. Other enzymes consume energy in different ways, but the production, maintenance, and operation of these molecular machines requires energy, because these machines do work. They keep the information in our cells safe from the ravages of entropy, just as a refrigerator keeps its insides cold despite Nature's attempts to bring it back to room temperature. Our cells are information-preserving engines, and they perform beautifully. Our genetic information remains virtually undisturbed after generations and generations of duplications.

In 1997, scientists got a graphic example of just how good our information-preserving machines are. A handful of biologists analyzed the mitochondrial DNA of a nine-thousand-year-old skeleton found in Cheddar, England. They extracted the genetic information from one of its molars and analyzed some fairly intact stretches of DNA. (Once the host organism dies, the information in it degrades owing to the ravages of entropy, but the pulp at the center of the molar had luckily stayed intact enough to yield some DNA samples.) The biologists analyzed a mitochondrial DNA segment that doesn't seem to code anything essential, so it should be mutation prone compared with the more essential parts of the genome. (That is, any mutation would not kill its host organism, so an error to that stretch of the DNA would not activate the ultimate error-checking mechanism, death.) But even though this was an error-prone region of mitochondrial DNA, when the scientists analyzed mitochondrial DNA samples from local Cheddar residents they found a nearly perfect match. Adrian Targett, a history teacher at a nearby school, had almost the exact same information in his mitochondrial DNA as was stored in the nine-thousand-year-old skeleton. In the stretch of four hundred As, Ts, Gs, and Cs that the biologists analyzed, Targett's mitochondrial DNA matched the skeleton's, symbol for symbol, except for a single mutation. There was only a two-bit difference in the eight hundred bits of information in the two men's mitochondrial DNA.

There is no way that near-perfect match can be a coincidence; the chances against it are astronomical. Perhaps Targett was a descendant of the skeleton's brother or sister; perhaps they were more distant relatives. But it is quite clear that even comparatively error-prone regions of our genome are very stable as the information duplicates itself over and over again. After nine thousand years of duplication, Targett was carrying almost the exact same sequence that the skeleton had.

More essential stretches of DNA—which kill an organism if they are tampered with—are conserved for even longer times. In May 2004, a group of scientists published a paper in *Science* that described five thousand relatively large sequences that appeared, 100 percent identical, in humans, rats, and mice; portions of these sequences are pretty much intact in the genomes of other mammals such as dogs, as well as other vertebrates, such as chickens and puffer fish. If, as scientists believe, this information was passed down from generation to generation from a single source rather than arising independently in these organisms, then the information must have been there before the mammal family tree split from the other vertebrates tens of millions of years ago, and even before fishes split from the branch that evolved into reptiles and birds hundreds of millions years ago. Through all that time, after billions of replications, the information remains more or less intact, surprisingly well protected from the ravages of time and entropy.

But this does not mean that our cells are exempt from the second law of thermodynamics. While our enzymes keep the cell's information safe—repairing it and reversing entropy locally—these proteins consume energy and do work. That means that the entropy of the universe must increase, even if the entropy of the cell is constantly kept low. (This is no different from the case of the refrigerator. Even though it keeps its belly cold by reducing its own entropy, it must expel heat and increase the entropy of the universe in the process.) In a sense, our cells are eating energy, and their waste product is entropy.

Luckily, our cells have a source of energy. The sun is the source of (most of) the energy available to creatures on Earth; it pours more than a million billion megawatt-hours per year on our planet in the form of light. Some organisms use that light directly, exploiting the energy in the photons to manufacture sugars out of carbon dioxide and water. Some use the light indirectly—by eating the organisms that use light directly. Or by eating the organisms that eat the organisms that use light directly. Or by eating the organisms that eat the organisms that eat the organisms . . . you get the idea.[6]

But what about entropy? Not only must organisms consume energy, they must discard entropy—or more precisely, they must somehow increase the entropy of their environment if they are to reverse the creeping degradation that the second law of thermodynamics exerts upon the information in their cells. Luckily for us, the Earth is a great place to dump entropy. It is a system way out of equilibrium, like a gas that is mostly on one side of a box.

If the Earth were a planet in equilibrium, it would look almost the same at every place on its surface. It would be at roughly the same temperature everywhere: the Sahara would be no different than the Arctic tundra. The atmosphere would have the same pressure everywhere: there would be no wind, no rain, no storms, no high and low pressure systems, no ocean waves, no warm days, no cool days, no polar caps, and no tropics. But that's not the Earth at all. Our planet is a dynamic place that changes day by day. The air pressure fluctuates as storm fronts move and air sloshes around the globe. Travel around the Earth and you will see very different environments: deserts, oceans, ice caps—places that are humid or dry, hot or cold, or all of them at different times of the year. This is not equilibrium, not by a long shot.

Since the Earth is out of equilibrium, there is plenty of room for us

6. A small number of organisms are not dependent on the sun as their source of energy. Certain creatures are able to use heat from the Earth's insides (which comes, nowadays, largely from radioactive decay of elements) and the chemicals that spew forth from the Earth's hot interior. It doesn't really matter *where* the energy comes from, but energy must be there in some usable form for life to exist.

to increase its entropy, moving it a touch closer to equilibrium. Humans, for example, consume energy in a quite accessible, usable form—such as Big Macs—but since energy cannot be created or destroyed, we are simply converting it into a less usable form such as waste heat (not to mention a brownish energy-containing product that is somewhat less appetizing than a Big Mac). We are constantly taking the usable energy from the sun and, directly or indirectly, making it less usable. In so doing, we are increasing the entropy of our environment—and our environment is the Earth. In time, if there were no way for the Earth to get rid of this entropy, our planet would slowly approach equilibrium. It would be harder and harder for organisms to shed their entropy by increasing the entropy of the environment, and life would slowly die out as Earth approached its state of maximum entropy. But this doesn't happen, thanks, again, to the sun.

If you observe the Earth from a distance, you will notice that it shines—not as brightly as the sun, to be sure, but it is radiating light. Some of that light is just a direct reflection from the sun, but some is not. Earth, as a system, absorbs light and reradiates it in an altered form. For example, the sun emits gamma rays, X-rays, and ultraviolet radiation that never reach the Earth's surface. These high-energy, high-temperature photons strike molecules in the atmosphere—such as ozone—and break these molecules apart. The energy of the photons breaks chemical bonds and makes atoms in the atmosphere move faster. It heats up the air above us. And hot things radiate energy in the form of photons.

However, the atmosphere is much cooler than the source of the X-rays, gamma rays, and ultraviolet radiation. Instead of radiating hot, high-temperature photons, it radiates cool, low-temperature photons, light like infrared radiation. Organisms help this process along, too: plants convert visible light into sugars, and animals convert plants into waste heat and infrared radiation. All in all, Earth's organisms convert visible light, which is created by objects that are thousands of degrees in temperature, to infrared light, which is created by objects at a few tens of degrees. The Earth and its creatures have converted hot photons into cool ones, and this cool infrared radiation streams out into space. This is a way of shedding entropy, reducing the entropy of the Earth at the expense of its surroundings.

Deep space is very cold. The background radiation that fills the universe is at about 3 degrees Celsius above absolute zero. If the universe as a whole were in equilibrium, its temperature would not be far above that. The entire cosmos would be only a few degrees from being as cold as physically possible. Anything that is hotter than that frigid equilibrium temperature, anything that is tens or hundreds or thousands of degrees above absolute zero, is not at the universe's equilibrium level. The hotter an object is, the more distant from the universe's equilibrium it is. And the more out of equilibrium an object is, the more entropy you can dump onto it, making it cooler and bringing it closer to universal equilibrium. That is precisely what the Earth and its inhabitants are doing. By taking the hot sun's energy, cooling it down, and reradiating it, our planet and the organisms that live on it are spitting entropy out into the solar system and beyond. It has taken a source of energy and made it less usable. In the thermodynamic units of entropy, the Earth is reducing its entropy by a little less than a trillion trillion joules per degree Celsius per year, sending all of it into the far reaches of space.

So, all told, the information in our cells is immortal because of this intricate exchange of energy, entropy, and information. Molecular machines in our bodies are following the instructions that our genetic information provides: they duplicate and maintain the information in our cells, consuming energy and creating entropy. They can do this because the organism itself gets energy, directly or indirectly, from the sun and releases entropy into the atmosphere or the sea—into the Earth's environment. The Earth sheds this entropy because of the sun's illumination. Energy flows in, entropy flows out, and the information in our cells is preserved.

This cycle can continue so long as the sun shines and the Earth exists. If the sun suddenly turned off, the Earth would quickly cool down. The oceans would freeze, the atmosphere would settle, and the whole planet would rapidly approach an equilibrium temperature a mere handful of degrees above absolute zero. All life would cease. But so long as there is a source of energy and a way to get rid of entropy, information can duplicate itself and keep itself relatively free of errors—and reverse the ravages of time. Information can be immortal, despite entropy's attempt to dissipate it.

Though scientists don't have a good answer for the question, What is life? this complex dance to duplicate and preserve information must be a major part of the answer. Information holds a good portion of the secret for understanding the nature of life. Not only that, it holds clues to another unanswered question: Where did we come from? Here, too, information is yielding surprising answers to this ancient puzzle.

The information in our cells is passed down from generation to generation, and written in our genetic code is our history as a species—our migrations, our battles—back to the very birth of humanity. And earlier still. Naturally, then, scientists can use information to look backward in time.

Deciphering a human's genome is like reading a long book written by all his ancestors. Each genome bears the signature of each of its predecessors, each genetic program that came before it in the chain of reproduction. Reading the information in each person's genome can reveal an interesting tale that is accessible in no other way.

One interesting example comes from Zimbabwe, where a tribe of people—the Lemba—tell a tale that's hard to believe. The legend, told by parents to their children countless times, tells of a man named Buba, who, three thousand years ago, led the Lemba southward from the lands that make up modern-day Israel. The Lemba claim to be a lost tribe of Judea: they claim to be Jews. After a long journey that took them through Yemen, through Somalia, and along the eastern coast of Africa, they finally settled in Zimbabwe.

Few believed the Lemba's story. There was little to link the tribe with the ancient Jewish people. It was true that, like Jews, the Lemba observe the Sabbath, refuse to eat pork, and circumcise their sons. On the other hand, oral traditions are unreliable, and hardly a reason to accept such an extraordinary claim of descent. Furthermore, the myth of the lost tribes of Israel is extremely common throughout the world; many peoples have claimed to be a lost tribe. Yet scientists found at least a grain of truth in the three-thousand-year-old legend, thanks to the information that the Lemba carry in their genes.

In 1998, geneticists in the United States, Israel, and England analyzed the Y chromosome of Lemba males. (The Y chromosome is the packet of genes that gives a male child his maleness. It is passed down from father to son to grandson. Females do not have a Y chromosome; they have a second copy of an X chromosome instead.) The Y chromosome is particularly interesting, as it can contain a strong marker of a people's Jewish heritage—priestly genes.

According to Jewish tradition, the priestly class, the *cohanim*, were a closely related group of people; in fact, according to legend, they all descended from a single male, Aaron, the brother of Moses. The title of priest, or *cohen*, was passed down from father to son to grandson to great-grandson since time immemorial. Just like the Y chromosome. Handing the office of cohen down from generation to generation was the same thing as handing down the Y chromosome from generation to generation. All Jewish priests, if the legend were true, should have the same Y chromosome: the one that Aaron himself had.

Reality isn't quite as simple as that. Not all cohen Y chromosomes are identical. But in 1997, scientists found a genetic marker of Jewish priesthood on the Y chromosome. They discovered that modern-day cohanim shared genetic characteristics that were fairly distinctive; even noncohen Jews did not have the same sorts of genes on their Y chromosomes. Because Jews hand down the priesthood along with the Y chromosome, every member of the priesthood had similar Y chromosome information, even though the Jewish population had scattered throughout the globe and mixed their genes with other peoples. Cohens faithfully passed down their distinctive genetic information from thousands and thousands of years ago, and all Jewish peoples that maintained their class of cohanim had a subpopulation of people who carried the markers of this priesthood. The Lemba were no exception. Even though they were separated from their Jewish roots, they had a cohen class that also shared the genetic information of the priesthood. These genetic markers indicated that the Lemba's priests were of the same stock as the other Jewish priests around the world—showing that the Lemba, too, had a Jewish heritage. It was a sure sign; the probability that they could have developed that particular marker on their own, through random mutation, is unspeakably tiny.

There were no written records of the Lemba's departure from Judea, yet their genes gave a more accurate picture of that migration than any historian ever could. Geneticists have used the information in our cells to reconstruct other human migrations as well. By comparing which populations share which distinctive genetic information—such as the genes for blood type—geneticists have been able to map how ancient peoples migrated, shifted, and interbred. They have also revealed how the human species nearly was destroyed.

In the late 1990s, geneticists at the University of California at San Diego analyzed the genetic diversity of different primates; that is, they saw how different the DNA sequences of individuals are from one another. Chimp and gorilla populations are genetically diverse—the mark of a large, healthy species—but the entire race of humans has less genetic diversity than an average group of a few score chimpanzees. What could have caused this incredible lack of genetic diversity?

If geneticists are correct, about 500,000 to 800,000 years ago something nearly wiped out our ancestors. Disease, warfare, or some other disaster destroyed most of the human population, dropping it down to a mere thousand or so individuals. That little group of humans managed to hang on, reproduce, and rebuild the species from their tiny number, but their descendants—us—have little genetic diversity. Our ancestors were forced through a genetic bottleneck; all of us are sons and daughters of this small group of primates. As a species, we humans are terribly inbred because of this disaster hundreds of thousands of years ago.[7] The only witness to this near catastrophe is the information within our genes.

This technique applies to the information not only within humans but within other species. By looking at how different the human genome is from the chimpanzee genome, and the chimpanzee genome is from the puffer fish genome, on down to the flatworm genome to the cyanobacterium genome, geneticists are able to reconstruct how the information propagated through the ages in organism after organism, back to times well before the chimpanzee's ancestors and our human ancestors parted ways roughly six million years ago. Scientists can trace information back through the age of the mammals, through the age of the dinosaurs, back almost to when the very first life floated in the primordial ooze of the ancient Earth.

The information in our genome was witness to the birth of life on Earth. It bears all the marks of its passage through the ages, all the scars of its evolutionary heritage.[8] It even may have signs of a time when the

7. Scientists are able to come up with dates for significant events in genetic history—such as a genetic bottleneck or the creation of a new branch on the tree of life—because the information in genes is equipped with a clock. Mutations. Though there is quite a lot of inherent uncertainty in the technique, and considerable controversy about how accurate these clocks are, scientists can get a rough estimate of how far back these events occurred by watching how mutations propagated through humanity's genes. If you have a handle on how often mutations occur, you can figure out how far back two peoples, or two species, split from each other. By comparing a similar stretch of information in the two groups' genomes and seeing how different the two genomes are—how many mutations have occurred since the two stretches were identical—you get a rough sense of when the split occurred.

8. Even though information is witness to evolution, creationists attempt to use information theory to *attack* evolution. Indeed, information theory is supposedly a bulwark of the "intelligent design" movement, but the information-theoretic arguments they

medium for carrying information in organisms wasn't DNA at all. Many scientists believe that at one point, the information of life might have been stored upon a related, but more fragile, molecule: RNA. Some biologists even believe that information was stored in a different medium before that. But whatever the original medium of life's information (and how that information first came to replicate itself), it is clear that the information of life has a history almost as old as our planet, and much of that history is preserved in every one of our cells.

Scientists don't know precisely how life began, but the near immortality of information has preserved a story that goes back to the very beginnings of life.

It's a grim picture: life might be nothing more than information's scheme to duplicate and preserve itself. But even if this is true, it doesn't provide the complete picture. Life is extremely complex, and our existence is not entirely determined by our genes. The environment also exerts its influence upon an organism's development—as does sheer dumb luck. And humans, more than any other species on the planet, have the ability to transcend the dictates of the information inscribed in every cell. We have spectacular brains.

We are beings who are able to communicate with and learn from one another. We can pass down knowledge from generation to generation and build upon it. With the help of centuries of work, scientists are on the brink of being able to alter our own genetic code, changing the information within us. How can we be slaves to information if we might soon be able to alter it at will?

Humanity is learning to understand and manipulate our genetic

---

make have severe flaws. For example, they argue that it is a violation of the laws of thermodynamics for a genome to gather more information over time, yet it is a fact that the sun's energy and Earth's entropy shedding allow organisms to preserve, duplicate, and modify their genomes, often increasing the amount of information that the genomes contain. Information theory doesn't undermine evolution; the situation is quite to the contrary.

code, but we do it *because* of information, not despite it. Our brains, as marvelous as they are, are machines for manipulating and storing information. Even so, for millennia, humans were unable to preserve that information against the damaging influence of time. Before there was a method of transmitting the information in our heads from person to person—language—and a way to preserve it against the distorting and damaging hands of time—writing—that knowledge was lost each time an individual died. The knowledge of a single human is not nearly enough to allow him to crack the genetic code. But language and writing allowed humanity to preserve its collective accumulated knowledge and preserve it even as all the individuals who gathered that knowledge died. And once the information was preserved, it was built upon by generations that followed. Only when humans came up with a method of transmitting and preserving information could they get on the path of defeating our mindless and implacable genetic program to transmit and preserve information.

Of course, this seems like a paradox, but it isn't. The information in our genes is of a very different sort from the information we process in our brains or the information we preserve in our language or writing.[9] But the same laws apply. Writing is a series of symbols—letters—that are reducible to bits; spoken language, too, is a series of auditory symbols—sounds known as phonemes—and they, too, are reducible to bits. Shannon's theory of information applies to writing and to language as it does to any string of bits. In fact, studies of writing and the even more ancient tool of language are yielding results similar to those of the genetic analyses of humans. (Unfortunately, language only reaches back tens of thousands of years instead of hundreds of millions.)

Take the Lemba, for example. The information in their language hinted at their Jewish heritage even before scientists were able to deci-

9. Though it need not be. If someone so desired, he could use a virus to insert, say, a passage from *Gulliver's Travels* into his genome, and it would be preserved for many, many generations.

pher the information in their genes. Though the Lemba spoke a Bantu language—a group of African languages that includes Swahili and Zulu—some of their words smacked of a foreign land. Some of their clans had Semitic-sounding names like "Sadiqui." (The word *sadiq* means "righteous" in Hebrew, and names like Sadiqui are found in the Jewish regions of Yemen.) Since language is a less reliable storage medium for information than our genome, the evidence of the Lemba's past is less apparent in their language than it is in their genes. But the evidence, the information about their ancestry, exists nonetheless.

There's evidence of other historical events as well. The information in language, like the information in our genome, shows the scars of major events in human history—of battles and invasions and migrations. The English language, for example, shows the marks of a foreign occupation. Until the eleventh century, Old English was purely a Germanic language. A word-order-preserving translation of the first sentence of the tenth-century poem "The Battle of Maldon" might go as follows:

> Commanded he his men each his horse to leave,
> fear to drive away and forth to go,
> to think to their hands and to courage good.

Notice how foreign this sentence structure feels. The verbs tend to be at the end of the sentence, rather than at the beginning. When compared to modern English ("He commanded each of his men to leave his horse, to drive away fear and to go forth"), tenth-century English sounds like a twisted mess. In fact, it is almost identical in structure to modern German, which often puts its verbs at the end of a sentence,[10] and is closer to modern German than it is to modern English.

---

10. Mark Twain offered this description in the late nineteenth century: "You observe how far that verb is from the reader's base of operations . . . ," he wrote. "Well, in a German newspaper they put their verb away over on the next page; and I have heard that sometimes after stringing along the exciting preliminaries and parentheses for a column or two, they get in a hurry and have to go to press without getting to the verb at all." Mark Twain, *A Tramp Abroad* (New York: Penguin, 1997), 392.

In 1066, a battle changed the evolution of the English language forever. The Duke of Normandy, William, successfully invaded England. A Frenchman, he subjugated the Anglo-Saxon kingdom, and soon his French-speaking comrades became the new nobility of England. The language of the court was French; the language of the peasantry was English. This out-of-equilibrium state did not last for very long, and as the French-speaking and English-speaking populations merged, so did their two languages. Within three centuries, the once-Germanic English assimilated a considerable amount of French grammar, including word order. (We tend to have the verb in the middle of the sentence, as in French, rather than at the end, as often happens in German.) English also adopted a great deal of French vocabulary, and a careful analysis using the French and German vocabularies alone can tell a linguist which side won the battle of Hastings. Look at the words for foodstuffs. *Beef* comes from a French word (*boeuf*), while *cow* comes from an Old English one. *Mutton* is French (*mouton*), while *sheep* is Old English. *Pork*, French (*porc*); *pig,* English. The English-speaking serfs, who lost the battle, tended animals. The French nobility, who won the battle, ate them. Our language is covered with thousand-year-old scars from the battle of Hastings. The information preserved in our language records our history, just as does the information in our genes.

Language and writing are one thing; our brains are another. It seems hard to believe that the information in our brains is similar to the information in our genes. For one thing, unlike our genetic information, which attempts to remain unchanged by the environment, our brains are constantly acquiring and adapting to information that they have gathered from the environment. The human brain is an information acquisition machine as well as an information-processing machine.

But the difference is academic, as far as information theory goes. Any information-processing machine must obey the laws of information theory. If the machine has a finite amount of memory (as our brains do), then it must expend energy in performing its calculations or it will grind to a halt. (As ours does. Though the brain constitutes only a few percent of an adult human's mass, it consumes about 20 percent of the energy we eat and the oxygen we breathe.) The information in our heads—and any signals in our brains, no matter how they are stored or transmitted—can be reduced to a series of bits and analyzed with Shannon's theory.[11]

It's a disturbing concept. From the perspective of information theory, the sloppy information-carrying circuits in the brain are no different from transistors or vacuum tubes or signal lamps or semaphore flags. They are the medium, not the message, and it is the message that counts. True, the brain is much, much more complicated than any other information-processing or storage device we know, but that complexity does not invalidate the laws of information. These rules apply to messages regardless of what form they are in. Though we only know a small amount about how the brain encodes and transmits information, and we know even less about how the brain processes it, we do know that this information follows Shannon's laws. And one of these laws is that the information is expressible in bits.

In a laboratory not far from Princeton, New Jersey, the biologist William Bialek has spent years trying to decipher the codes that animal brains use to encode information—with some success. Most of his work has to do with flies. In experiments reminiscent of a teeny-tiny version of *A Clockwork Orange,* Bialek immobilizes the flies, sticks needles into their optic nerves, and forces them to watch movies. But these gruesome-sounding experiments have a point. Bialek and his colleagues have been recording the signals in each fly's brain when the fly sees different things, and this, in turn, reveals how information is encoded in the brain.

Fly brains, like human brains, are made up of specialized cells known as neurons. These neurons are connected to one another in an enormous network. If you tickle one of these neurons just right, it will

11. There is one possible exception to this, which will be described later in the book: that the information in our heads is *quantum* information rather than *classical* information.

"fire." Through a complicated electrochemical process, sodium and potassium ions on opposite sides of the cell membrane switch places. The neuron goes from **0** to **1**, and then after a small fraction of a second it switches back, reverting to a **0** again. Though neurons have intricate systems for messaging each other, for turning up and down the volume on their input and output ends, the firing of the neuron is essentially all or nothing: it fires or it doesn't. It's pretty much a binary decision, and you don't even need to invoke the more advanced Shannon argument to imagine that neural signals can be reduced to bits and bytes.[12] A neuron is apparently a classical channel for information.

Bialek has been trying to figure out how the fly encodes messages on that channel. With the sensors placed in a fly's optic nerves, Bialek shows it movies of very primitive images: a white bar, a dark bar, a bar moving from left to right, and so forth. He records what signals are passing through the optic nerve to the brain. By deciphering those signals, Bialek has been figuring out the basic "alphabet" of neural signals that the fly's brain uses to encode visual information. And he's been figuring out how much information those signals encode. Though there is some argument about the exact numbers, a neuron in a fly's brain seems to be able to transmit, at its peak, about five bits of information per millisecond. Bialek's work is confirming that even something as complicated as a visual image on a retina is reduced to the equivalent of bits and bytes and transmitted into the brain. When a fly sees a tasty chunk of potato salad and decides to approach, its brain is simply receiving a string of bits of information from its eyes, processing those bits, and sending signals to its muscles—also quantifiable in bits—to fly toward the food. Even though it is an extraordinarily complex bit-processing machine, the fly's brain is a bit-processing

---

12. The conversion to bits is not as straightforward as it looks on the surface, though. A neuron's signals are 0s and 1s, but the coding scheme in the brain makes use of the *timing* of those 0s and 1s rather than simply treating them as a string of bits. Nevertheless, Shannon's theory says that this code, as complicated as it might be, is reducible to a string of bits.

machine nonetheless. And, according to classical information theory, so is ours.

This is an even grimmer picture than before. Even though we are able to pass information from generation to generation and use our brains to create things as sublime as the *Odyssey* and as fascinating as quantum field theory, as far as scientists can tell we are pretty much information-processing machines. Incredibly complex information-processing machines, ones capable of tasks that no other such machine is capable of, but information-processing machines nonetheless.

There seems to be something missing in this picture. After all, we are intelligent, sentient, self-aware beings. We are conscious—and other, nonliving information-processing machines, such as computers, don't seem to be. What is it that separates us from calculators and computers? Is it mere scale, or is there something else at play?

Some scientists and philosophers (not to mention religious leaders) think that there is. However, if you concede that information is what's being transferred in our neurons, there's hardly a way to escape the dark and reductivist conclusions of classical information theory.

But there might be a way out. Information theory, as envisioned by Shannon, is not complete. While it describes the information that can be stored or transmitted by computers and telephones, by telephone wires and fiber-optic cables, the laws of information theory are predicated on classical physics. And in the twentieth century, two revolutions ended the classical era of physics: relativity and quantum theory.

Relativity theory and quantum theory changed the way physicists perceived the universe. They banished the naive, commonsensical, mechanical universe and replaced it with one that is much more intricate and much more disturbing, philosophically. At the same time, relativity and quantum theory altered the discipline of information theory just as they altered the rest of physics. Relativity, which describes mind-bending effects that happen when objects go very fast or are subjected to intense gravitational fields, put a limit on how quickly information can be transmitted from place to place. Quantum theory, which deals with the counterintuitive properties of very small objects, showed that there is more to information—at least in the subatomic realm—than bits and bytes. Yet, at the same time, information theory has altered those two revolutions in ways that scientists are only just beginning to understand. By looking at relativity and quantum theory in information-theoretic terms, physicists are getting the keys to the most important problems in science. But to get at those answers, we have to delve into both quantum theory and relativity—both of which are, fundamentally, theories of information.

FASTER THAN LIGHT

There was a young lady named Bright,
Whose speed was far faster than light;
  She set out one day,
  In a relative way,
And returned on the previous night.

       —A. H. Reginald Buller,
       Relativity

Just as the classical era ended with the fall of Rome in the late fifth century, so, too, did the era of "classical" physics end with the development of the theories of quantum mechanics and relativity in the early twentieth century. At first glance, neither of these revolutions—each of which involved a young scientist named Albert Einstein—involves information. But looks are deceptive.

Even though relativity and quantum mechanics came before Shannon's theory, both are actually theories of information. It is a little tricky to see at first, but the fundamentals of information theory lie underneath the surface of both these theories. And information theory may well be the key to unraveling the mysteries of relativity and quantum mechanics, and the troubling conflict between them. If it does, it will be the crowning triumph of modern physics; scientists might well have a "theory of everything," a set of mathematical equations that describes the behavior of all objects in the universe, from the smallest subatomic particles to the largest galaxy clusters. The revolution begun in a quest to figure out how many telephone calls can fit on a copper cable may well lead to a fundamental understanding of every object in the cosmos.

To understand how information can have such broad and deep importance, we have to go beyond the thermodynamics and Shannon's "classical" theory of information. We have to explore the realms of both relativity and quantum theory, and this will lead to scientists' current understanding of information—and how it shapes the universe.

Both quantum theory and relativity are intimately related to entropy and information. Albert Einstein, who sparked both the quantum and relativity revolutions did so in part because of his earlier interest in entropy, thermodynamics, and statistical mechanics. Indeed, the first of Einstein's revolutions, relativity, is a theory that is directly concerned with the exchange of information: its central idea is that information cannot travel faster than the speed of light. Nevertheless, this doesn't stop physicists from building faster-than-light devices and time machines. Some of them actually work.

Einstein was an unlikely figure to revolutionize physics—but not quite as unlikely as some writers make him out to be. Contrary to legend, he never failed mathematics in school; all accounts paint him as a gifted math student. And though he was merely a lowly patent clerk, Einstein happened to be a clerk with a degree in mathematical physics. (His narrow-minded physics professor gave all his other classmates assistant professorships, but thanks to a personality conflict, Einstein was left without a university position when he got his degree.)

After searching for a university position and working briefly as a substitute teacher, Einstein took the job at a patent office in 1902 to make ends meet. It was a wise thing to do, as he was married within a year, and shortly thereafter he was a father. Even as he labored in the patent office, though, he was not merely a lowly patent clerk. He was a trained physicist nearing the peak of his powers, and he completed his dissertation and published a large number of scientific papers in a very short time. It would be a few years yet before he would formulate the theory that made him famous; from 1902 through 1905, he was obsessed with another area of physics altogether: thermodynamics and statistical mechanics—the stuff of Boltzmann.

In 1902, Einstein published a paper on entropy in the *Annalen der Physik,* and in the next year he followed up with one about reversible and irreversible processes. In 1904, he wrote a paper about measuring Boltzmann's constant, the k that appears in his entropy equation. None of these papers was terribly influential, in part because he wasn't fully familiar with all of Boltzmann's writings. Einstein also investigated the implications of the statistical, random motion of matter: he studied Brownian motion, and his doctoral thesis had to do with using statistical methods to determine the radii of molecules. These studies would soon come to an end, for Einstein was on the brink of his miraculous year, 1905, when he would turn to more important work and turn physics on its head.

Einstein's fame comes from his theory of relativity. One of his crucial papers in 1905 was a limited version of relativity. This first version did not work under all conditions, though; it did not apply when objects were accelerating or feeling the pull of a gravitational field, for example. But this paper, which introduced the *special theory of relativity,* was simple, profound, and correct. It solved a lingering problem that had troubled physicists for decades, one that seemed unrelated to the problems of information and thermodynamics. Nevertheless, the solution to this problem turns out to be information-theoretic: Einstein's theory of relativity, at its heart, is a theory of how information can be transferred from place to place. But to get to that understanding, we have to go back to the heart of the problem, which was discovered long before scientists started thinking in terms of information.

The problem: light was misbehaving. This was a serious dilemma, since physicists thought they had figured out the fundamental properties of light in the early and mid-1800s. They thought they knew what light was, and they thought they understood the equations that governed how light behaves. Physicists were wrong on both counts.

A heavy debate about light went back for centuries before Einstein was born. Isaac Newton, the founder of modern physics, was convinced that light was a collection of tiny particles that traveled instantly from place to place. Other scientists, like Christiaan Huygens, the inventor of the pendulum clock, argued that light was not a particle at all; it was more like a wave of water than a single, discrete object. The arguments rattled back and forth, with most physicists leaning toward Newton's idea that light was a corpuscle—a particle—but it was really a matter of faith whether one believed light was a particle or a wave. Nobody could come up with a definitive experiment that would distinguish which side was right. Nobody, that is, until 1801, when the British doctor and physicist Thomas Young devised an experiment that, to all appearances, answered the question and settled the matter once and for all.

Young's experiment was very simple. He shined a beam of light through a barrier that had two narrow slits. On the far side of the barrier, the light created a pattern of fine light and dark bands—an interference pattern. These fringes were very familiar to those who studied waves.

Interference patterns are created by waves of all sorts. You've probably seen them before, even if you weren't aware of the phenomenon. When you drop a stone into a lake, you create circular ripples in the water. The splash of the rock as it hits the surface of the pond sets up a series of alternating crests and troughs that spread quickly away in all directions. The rock makes a circular pattern of waves. Now, if instead of a single rock, you drop two stones next to each other at the same moment, the pattern is much more complicated. Each stone sets up its own pattern of crests and troughs. These crests and troughs spread out and they run into each other—and interfere. When a crest runs into a trough, or a trough runs into a crest, the two ripples cancel each other out, leaving a perfectly calm surface in their place. If you drop two stones into a calm lake, you might even be able to see where the rippling surface is scarred with lines of still, calm water. These

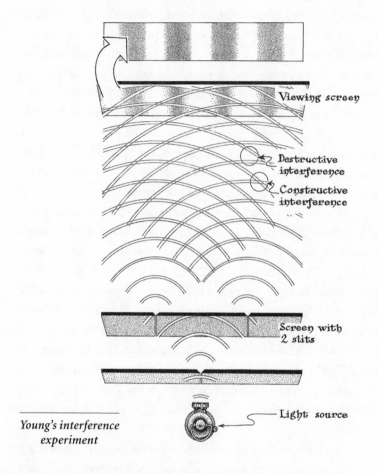

Young's interference
experiment

Viewing screen

Destructive
interference

Constructive
interference

Screen with
2 slits

Light source

lines are regions where the crests from one stone always cancel the troughs from the other stone and vice versa. The lines are an interference pattern, precisely what Young was seeing in his experiment with light.

Light going through two slits in a wall is just like two stones hitting the water at the same time. Just as would be the case with water waves, light's crests and troughs pass through the slits and then stream rapidly away from the barrier. Just as with ripples on the surface of a pond, each crest or trough of light that goes through the left slit is constantly running into crests and troughs that have gone through the right slit. When crest meets crest or trough meets trough, the two reinforce each other; however, when crest meets trough, the two cancel. Viewed from above, the regions of cancellation make a pattern of dark stripes—where light cancels itself out—identical to the stripy pattern of calm water created by two stones dropped in a lake. But with light, you're unable to see these lines from above; you project them on a screen at the far end of the room. Young saw that when light hits a screen after passing through two slits, it leaves a series of light and dark stripes. It creates an interference pattern.[1]

Young's discovery—the detection of interference patterns in light—showed that light behaved like a wave, as interference is inherently a wavelike property. Physicists could not explain interference patterns by means of particles that collide and bounce, but it was easy to explain them, in great detail, via waves that pass through and interfere with each other. Light seemed to be a wave, not a particle, and Young performed a number of other experiments that reinforced this notion. He saw that light does other wavelike things, such as diffracting. It bends slightly when it hits a sharp edge—it diffracts—something that waves tend to do and corpuscles tend not to do. The verdict seemed pretty clear to the physicists of the time: light was a wave and not a particle.

The light-wave case slammed shut in the 1860s when James Clerk Maxwell—he of the demon—derived a set of equations that explained how electric and magnetic fields behaved. Light, which is an electro-magnetic phenomenon, follows the rules set down in these equations,

---

1. The easiest way to make a nice interference pattern is to shine a laser pointer parallel to a mirror in the bathroom—make a spot on the wall perpendicular to the mirror. When you look in the mirror at the reflection of that spot, you will see a pattern of bright and dark lines, an easy-to-see interference pattern. This pattern is caused by phenomena a little more complicated than the two-slit phenomenon: it is due to the laser light's bouncing off the mirror and interfering with the laser light that bounces off the glass that covers the mirror. Nevertheless, the principle is the same as the two-slit experiment.

too. To mathematicians and physicists, Maxwell's equations looked very similar to the equations that describe how waves propagate through a medium: they were wave-ish, and they described the way that light moves with great precision. In fact, the equations dictate the speed of light; apply Maxwell in the right way and you know precisely how fast light goes. The argument was overwhelming to the physicists of the nineteenth century. Light was a wave. But a wave of what?

When you hear a sound wave, you are listening to a splash of air knocking into your eardrum. When you clap your hands, you knock air molecules that knock other air molecules that knock other air molecules. This sloshing of air is the sound wave, which propagates toward your ear and makes your eardrum wiggle. Similarly, a water wave is the sloshing of water; molecules of water jostle each other as a crest or a trough speeds toward shore. In each case, the individual molecules in the wave don't move very far; they wiggle around a little bit. The overall pattern in the medium—water or air—can travel great distances, and it is that pattern that makes up the wave.

If light is a wave, what is getting jostled? What's the medium that light propagates through? In the nineteenth century, physicists had little idea of what this medium could be, though they agreed that it must exist. They dubbed this hypothetical medium, the carrier of light waves, the luminiferous ether.

In 1887, two American physicists, Albert Michelson and Edward Morley, tried to detect this ether with a technique that exploited the motion of the Earth. As our planet moves around the sun, and the sun moves around the center of our galaxy, the Earth should be hurtling through this ether like a speedboat over the surface of the ocean. This means that the Earth should be buffeted by an ether "wind" that changes velocity as the Earth orbits the sun. Thus, a beam of light going upwind should be moving at a different speed from a beam of light going downwind or across the path of the wind. Michelson and Morley therefore reasoned that if they would send beams of light in different directions relative to the ether wind, the two should travel at different speeds.

The two set up a very clever experiment to find this speed difference. At its heart was a device now known as the Michelson interferometer, which exploits the wavelike nature of light to make very precise measurements of distance or speed. The interferometer splits a beam of light and sends it down two different paths of the same size. When a crest of the wave hits the beam splitter, it divides into two crests, which then zoom in different directions, bounce off mirrors, and are recombined at a detector—perhaps a screen. Since the paths are the same size, the crests should arrive at the same time—if both beams move at the same speed. Crest will reinforce crest making one big crest, and experimenters would see a bright spot on the screen where the beams recombine. If, on the other hand, the ether wind retards one of the beams relative to the other, then one crest will be delayed. In fact, if the instrument is set up in the right manner, then the crest will arrive from one beam at precisely the time a trough arrives from the other. When the two beams are recombined, instead of reinforcing each other, crest to crest, they cancel each other, crest to trough, and the bright beam becomes a dark spot. So, with a Michelson interferometer physicists could detect the subtle effect of the ether wind. All they had to do was measure how changing their apparatus's orientation to the wind caused the bright spot to appear or disappear.[2]

However, no matter how the two experimenters tried, the speed of light was the same in every direction—whether the light was going upwind or downwind or sideways. In 1904, Morley even tried the experiment atop a hill to make sure that the laboratory wasn't somehow shielding the interferometer from the ether wind. It made no difference. The speed of light was the same in every direction, regardless

2. Since modern physicists know that the speed of light is a constant, they use a Michelson interferometer to measure distance rather than speed. If the two arms are of slightly different length, then you can get a dark spot rather than a bright spot.

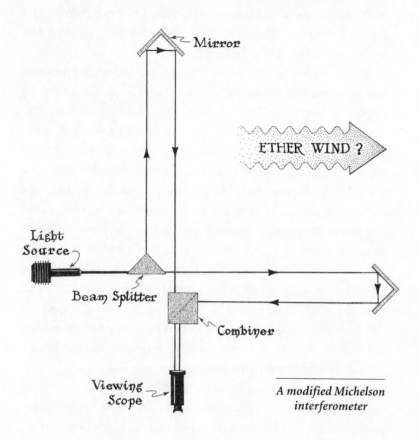

A modified Michelson interferometer

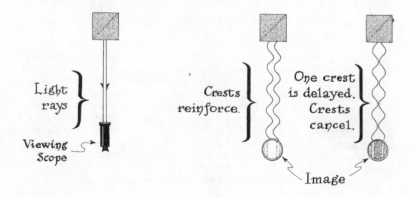

of the Earth's motion. There was no ether.[3] The Michelson-Morley experiment exposed a huge problem with the ether theory, and it won Michelson the 1907 Nobel Prize in Physics.[4]

That was half the problem with light—explaining light propagation without an ether medium to carry the wave—but there was yet another sticking point. It was with Maxwell's equations. These equations did an absolutely spectacular job of describing how electrical and magnetic fields—and light—behave. They were, arguably, the biggest triumph of nineteenth-century physics; they wrapped up the mysterious nature of electromagnetic fields in a nice little gift-wrapped package with a bow on top. Unfortunately, there was one flaw. Move a little bit and the equations broke down completely. More precisely, Maxwell's equations only held for an observer who was stationary. If someone were on a train passing by the experiment and tried to describe the experiment from his point of view, from his "frame of reference," he would be unable to do so with Maxwell's theory. Maxwell's equations simply did not work from a moving frame of reference: the electric fields started turning into magnetic fields and vice versa, and when a moving observer totaled up the forces that were acting on a particle, the observer would often get the wrong answer. A trainbound physicist might calculate that a particle moves up into the sky, while a stationary physicist might conclude that the particle goes down into the ground.

This made no sense. The same laws of nature should apply no matter how an observer moves. An observer who is moving in a train and uses Maxwell's equations to figure out how a particle behaves should

3. There was another experiment that, in retrospect, seemed to belie the idea of an ether. In the mid-nineteenth century, the French physicist Armand Fizeau measured the speed of light in moving streams of water, expecting to see the ether dragged along with the water. He didn't see any such effect. In fact, it seems that Einstein was more influenced by Fizeau's experiments and observations of how the apparent positions of stars in the sky change depending on the Earth's orbit—a phenomenon known as stellar aberration that is due to the finite speed of light—than he was by the Michelson-Morley experiment.
4. Interestingly, in school Einstein was ignorant of the Michelson-Morley experiment and proposed doing a similar test of the ether. The aforementioned narrow-minded teacher, Heinrich Weber, refused to allow the young Einstein to do the experiment. Weber apparently didn't think much of the newfangled physics of the day.

get the same answer as an observer who is standing still. In fact, this idea—that the laws of physics do not depend on the observer's motion—is the first of two key assumptions of relativity theory. In 1905, Einstein stated that this "principle of relativity," which gave its name to the theory, *must* be true. The laws of nature cannot depend on an observer's motion. And though the principle of relativity is rather straightforward, it is subtle; it takes a little work to see how the universe could be any other way. Einstein's second assumption, on the other hand, was as subtle as a sledgehammer.

Michelson and Morley showed that the speed of light was unaffected by the Earth's motion. Einstein assumed that the speed of light was unaffected by *any* motion, instantly explaining the Michelson-Morley experiment. No matter how you are moving, you will always measure a light beam as moving at 300,000,000 meters per second: *c*, the speed of light. However, on its face, this assumption seems utterly absurd.

If you're walking down the street and a fly suddenly smacks you in the nose, you're barely going to flinch. A tiny fly can only flutter about at a few miles an hour, and with its tiny mass, no matter what the circumstances, it can only have a minuscule impact if it smacks into you. But when you drive down the highway on a warm summer's day, you'll occasionally hear a good, solid "thwack" as some hapless fly smears itself over your windshield. At highway speeds, if that fly hits you in the face, it might do some damage. It might knock off your glasses or even give you a bloody nose. That's because your motion affects how you perceive the speed of the fly: the relative motion of the fly is very different when you are standing still than when you are zooming by in a car. If a fly moves at 10 miles an hour, when you are at rest, it will smack into you at 10 miles an hour. This means that it has very little impact. However, if the fly hits you when you are moving 80 miles an hour, to your nose it seems as if the fly is moving $80 + 10 = 90$ miles an hour, and it will result in a much greater splat. In classical physics, and in our everyday commonsense world, speeds are additive. If you move relative to an object, you add your velocity to its velocity, and that is how fast it seems to be moving from your perspective. Everything in the world that we are used to works like this.

A policeman with a radar gun, for example, has to take his own velocity into account when tracking a speeder. A radar gun that tracks a speeding car moving at 100 miles an hour relative to the ground will give a different reading if the gun is moving. If the officer is stationary, his radar will obviously get a speed of 100 miles an hour when he zaps the car. However, if he is moving along with traffic at, say, 60 miles an hour, the radar gun will only see the 40-mile-per-hour difference in speed between the officer and the speeder. From the moving patrol car's point of view, the speeder is only moving away at 40 miles an hour. Conversely, if the officer is traveling in the opposite direction at 60 miles an hour, the radar gun will show an impressive 160 miles an hour. The speeder is moving at 160 miles an hour relative to the patrol car, even though the lawbreaker is moving only 100 miles an hour relative to the ground. What the radar gun displays when it measures the velocity of the speeder's car depends on how the policeman is moving: the outcome of the measurement depends on the officer's frame of reference.

If you replace the speeder with a beam of light, Einstein's constant-speed-of-light hypothesis is roughly equivalent to saying that the speeder is *always* clocked at 100 miles an hour, no matter how the officer is moving. A stationary policeman would see the speeder approach at 100 miles an hour and then zoom away at 100 miles an hour. An officer moving in the same direction as the speeder would still see the speeder approach at 100 miles an hour and then shoot into the distance at 100 miles an hour. An officer moving toward the speeder would also see the speeder zoom toward him at 100 miles an hour and away at 100 miles an hour. It is as if the speeder totally ignores the motion of the policeman. Obviously, this does not happen in real life, otherwise none of us would ever get speeding tickets—a radar gun would be totally unreliable!

Taken together with the principle of relativity, the constant-speed-of-light hypothesis seems untenable. If you had three cops, all moving

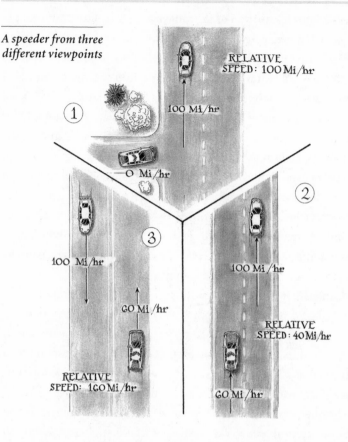

A speeder from three different viewpoints

in different ways, measuring the same light beam at the same time, the constant-speed-of-light hypothesis states that, despite their very different motions, they all measure the same speed for the light beam: 300,000,000 meters per second. How can three observers moving in different ways come up with the same measurements and at the same time, according to the principle of relativity, all be correct? It seems impossible.

But it is possible, and it is consistent. The three observers all do measure the same speed for light, and they are all correct. Modern measurements bear this out to great precision. No matter whether a satellite is approaching you or receding away from you, its signal always zooms toward you at the same speed: 300,000,000 meters per second. How, then, to get around the contradiction?

The answer lies in the concept of speed—and information. Speed is simply the distance traveled over a given amount of time. But you can't miraculously intuit how fast something is moving; you have to measure its speed somehow. You have to gather information about distance and time—say, you watch how far the object moves (using a meterstick) in one second (using a clock). If three police officers are measuring a light beam's speed, they each, independently, are effectively gathering information about time and distance with respect to their own metersticks and clocks. The only way out of the seeming contradiction caused by Einstein's two hypotheses is to assume that the clocks and the metersticks are affected by motion. This throws out millennia-old assumptions about time and distance. No longer can they be considered to be fixed, unchanging, objective quantities. Time and distance are relative. They change, depending on your frame of reference. And when your concepts of time and distance change, they affect how you measure speeds.

Back to the speeder. Assume, for the moment, that the speeder is a beam of light. Three supercops, moving in different ways (say, one stationary, and two moving in opposite directions at three-fifths the speed of light, or $0.6c$), measure the velocity of the speeder, and all come up with the same answer: the speed of light, c, is 300,000,000 meters per second. Why is this? Because each cop has to measure time and distance, and their metersticks and clocks are all messed up. When the stationary cop looks at his meterstick, he sees that it is the normal length; he listens to his clock, and it is ticking away at the usual rate. However, if he looks at the cops who are each moving at $0.6c$, he sees that their metersticks have shrunk by 20 percent: each is 80 centimeters long instead of the full 100! Furthermore, he sees that the two moving cops' clocks have slowed down. When the stationary officer counts off ten seconds by his clock, he notices that the moving cops' clocks have each only ticked off eight seconds.

"Aha! Here's the problem," thinks the stationary cop. "When I

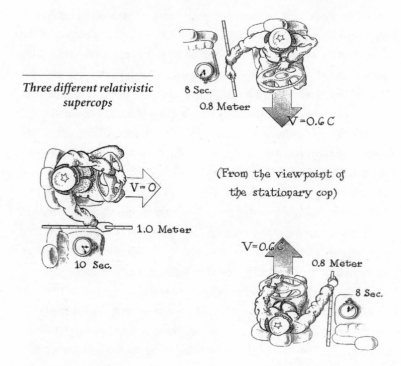

Three different relativistic supercops

8 Sec.

0.8 Meter

V=0.6 C

V=0

(From the viewpoint of the stationary cop)

1.0 Meter

10 Sec.

V=0.6 C

0.8 Meter

8 Sec.

measure the speed of light, I get the right answer because my meterstick and clock are working correctly. But the two moving cops get incorrect measurements because their sense of distance and time is distorted." It turns out that this distortion of space and time—the perception that the moving cops' clocks are slow and their metersticks short—brings all three measurements into agreement: the stationary cop measures the speeder moving at *c* as does each of the two moving cops with their short metersticks and slow clocks.[5] So, from the stationary cop's point of view, the two moving cops get the right answer, *c*, but only after the distortion of their metersticks and clocks is factored in.

Strangely, neither of the moving cops notices his meterstick shrink or

5. The numbers *do* work out, even though it is not obvious. The mathematics used for swapping perspectives is known as the Lorentz transformation, and it is slightly more complicated than the simple addition of the everyday speed conversions we're used to.

his clock slow down. In fact, when each moving cop looks at his meter-stick and clock, everything looks normal, but when each looks at the *other* cops' metersticks and clocks, he sees that the metersticks have shrunk and the clocks have slowed down. So each of the moving cops thinks, "Aha! Here's the problem!" and blames his colleagues' messed-up metersticks and clocks for getting the right answer in the wrong way.

Shrinking metersticks? Slowing clocks? It seems silly, but it has been observed. For example, particle physicists see clocks slowing all the time. Certain subatomic particles, like the muon or the tau parti-cle, heavier siblings of the electron, only have a short time to live before they spontaneously decay into other, more stable particles. (The muon, for example, lives, on average, about two-millionths of a second.) In a particle accelerator, though, a muon often travels at more than 99 percent of the speed of light, and as a result its internal clock is slowed relative to the laboratory's clock. This means that the muon lives a lot longer than it would if it were at rest. Global Position-ing System receivers, which sense clock signals from satellites orbiting the Earth, have to take into account relativistic clock slowing when fig-uring out position. Even more directly, in 1971 two scientists flew four atomic clocks aboard commercial jetliners. Because of their motion relative to the Earth, the clocks disagreed after the trip. Length con-traction and time dilation, as well as other strange relativistic effects, such as an increase in mass at high speeds, are a fact. They have been observed, and they agree wonderfully with Einstein's theory.

Einstein's two assumptions, the principle of relativity and a constant speed of light, had lots of weird consequences, but there is a beautiful symmetry to the theory. Observers might have very different views of the world—they might disagree about length, time, mass, and many other fundamental things—but at the same time, all the observers are correct.

In other words, Einstein's theory, at its root, says that you can-not divorce perception—the information you gather from your environment—from reality. If an observer collects accurate informa-tion about something (how fast a speeder is moving, for example), that information will be correct, but with a catch: it is correct only from his point of view. Different observers, making the same measurement and gathering the same information, will often get different answers. They may all get different numbers for how fast an object is moving, how long an object is, how much it weighs, or how fast its clock is ticking. However, no observer's information is more or less correct than any of the other observers' information. Everyone's information is equally correct, even though the answers to the questions about mass, length, speed, and time seem to contradict one another. It seems hard to accept, but the equations of general relativity work out beautifully. If you know how each observer is moving, you can use the equations to predict exactly what each observer sees; in other words, you are able to take the information you gathered and use the equations to figure out what the other observers are seeing. This is the key to understanding relativity. Different observers can ask the same questions about the same phenomena and get seemingly different answers. But the laws of relativity govern the laws of how information is transferred from observer to observer and tell you how different observers will interpret the same phenomenon in different ways.

The elegant way the equations worked out, not to mention the observations that they explained, convinced physicists that Einstein was correct. In the early 1920s, a rumor spread that a more sensitive Michelson-Morley–type experiment had detected the faint hints of a luminiferous ether, thereby disproving the theory of relativity. Einstein's famous response was, "Subtle is the lord, but malicious he is not." Einstein, like many other physicists of the day, was absolutely convinced that the theory was right. Relativity was too beautiful to be wrong.

However, there is one thing that physicists enjoy more than building a beautiful theory—and that's smashing someone else's beautiful theory. Plenty of people tried to destroy Einstein's. Since experimental tests of relativity are hard to do (and some predictions of general relativity haven't yet been tested because of that difficulty), theorists attacked Einstein's theory with a different tool: the thought experiment.

In a thought experiment, a physicist sets up a scenario and tries to solve it using the laws of the theory he is testing. If the theory has a hole and if the physicist is clever enough, he can set up a scenario that causes an internal contradiction, a point where the theory disagrees with itself. If this happens, if the theory is inconsistent, then it must be wrong. If the theory is sound, however, the seemingly paradoxical scenario will have a consistent explanation, and everything works out in the end. (Maxwell's demon was essentially a thought experiment, and it caused no end of problems for thermodynamics.)

Einstein himself loved thought experiments and used them to try to tear down other people's theories (as the next chapter will demonstrate). With relativity, the situation was reversed. Einstein had to contend with other scientists' thought experiments. One of the most tricky was what we will call the spear-in-the-barn paradox.

Imagine a sprinter with a fifteen-meter-long spear. He runs toward a fifteen-meter-long barn with two doors—a front door and a back door. To start off with, the front door is open and the back door is shut.

Now this sprinter is really good. In fact, he can sprint at 80 percent of the speed of light, and he runs into the barn. From the point of view of a stationary observer sitting in the rafters, the sprinter's spear is contracted (because of the relativistic effect on the runner's meterstick). In fact, the fifteen-meter spear is only nine meters long. If the observer in the rafters were to take a snapshot of the spear or measure it in some other way, he would see that it is only nine meters in length, even while the stationary barn stays at its original size of fifteen meters.

In other words, if a stationary observer tries to get information about the length of the spear, he will discover that it is nine meters long. And as Einstein's theory says, information is reality. If your (accurate) measuring instrument gathers information about the spear and that information reveals that the spear is nine meters long, then it is nine meters long—never mind that it started off as a fifteen-meter-long spear.

A nine-meter-long spear fits nicely in the fifteen-meter-long barn;

an electronic sensor can shut the front door as soon as the spear is fully inside the barn. For a moment, the spear is entirely enclosed in the barn, which has both doors shut. Then, just as the tip of the spear reaches the end of the barn, another sensor opens the rear door, letting the sprinter out. So far, so good.

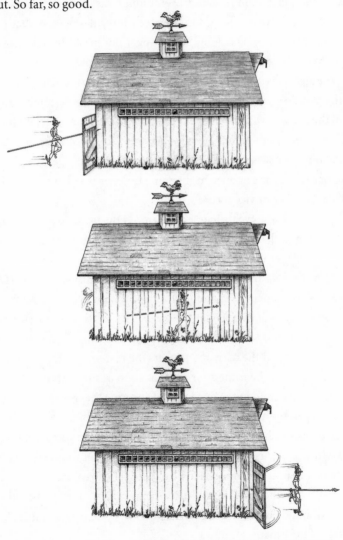

The spear-in-the-barn paradox from a stationary point of view

But things really get weird when you look at the events from the sprinter's point of view. From his perspective, the barn is rushing at him at 80 percent of the speed of light. If he were to gather information about how long the barn is, he would see that it's only nine meters long—and perception is reality. Even though his spear appears to be the full fifteen meters long, the sprinter's information says that the *barn* is only nine meters long, so the spear doesn't fit into the barn! How, then, could both doors be shut at the same time?

The answer is hidden in the last word of the question. The solution to the paradox has to do with time, but it is a little more complicated than the mere slowing of a clock. One of the side effects of relativity is that the concept of simultaneity—that two things can happen at the same time—breaks down. Different observers can disagree about whether two events happen at the same time, or whether one occurs before the other or vice versa.

In this case, the events in question are (1) the front door's shutting and (2) the back door's opening. From the point of view of the stationary observer in the rafters, the sprinter runs into the barn, (1) the front sensor shuts the front door, with the sprinter inside, and then (2) the rear sensor opens the back door, letting the sprinter out. But from the point of view of the sprinter, the order of events is reversed. He runs into the barn and (2) the back door opens when the tip of the spear reaches the end of the barn and triggers the rear sensor. He continues on, and then (1) the front door shuts as soon as the butt of his spear passes the threshold of the front door, triggering the front sensor.

The sprinter and the observer in the rafters disagree about the order of events, but mathematically the two observations are consistent with each other. The two sensors are independent, and there is no particular reason why one has to be triggered before the other. In one frame of reference, the front sensor triggers first, and in the other frame of reference the rear sensor triggers first. Once again, it is all a matter of information transfer.

Information doesn't get from place to place instantly; at most, it

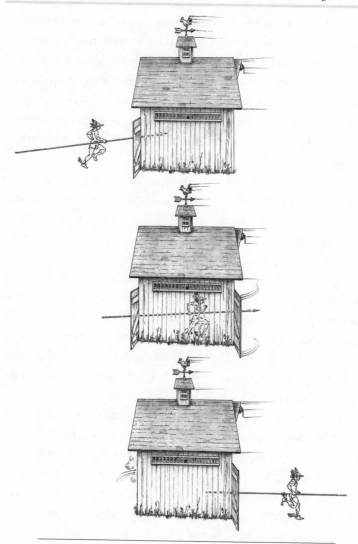

The spear-in-the-barn paradox from a moving point of view

can travel at the speed of light. This means that the concept of "simultaneous" doesn't really mean anything, because you have to take into account the fact that it takes time for information to travel to your observers. And an observer's motion will affect the order in which information reaches him. The information that the front door is shut and the information that the back door is open might reach one observer at the same time; to another observer, the "front door is shut" information might arrive first. To yet another, the information that the back door is open might come first. The three observers will disagree about whether the front door shut first, the back door opened first, or both events occurred at the same time. Which one is correct? They all are.

Einstein's theory of relativity says that an event only "occurs" from your perspective when the information about that event's occurrence reaches you. An event doesn't *really* happen until that information (traveling at the speed of light) traverses the distance from the event to you. Once again, perception—and information—is reality. This is what causes simultaneity to break down; since the three observers get information in a different order, then in truth the events they were observing *occur* in a different order for each of the three observers. It's a strange concept, but the breakdown of simultaneity in relativity theory is just something that physicists have come to live with; it doesn't violate any principles any more than do length contraction and time dilation. Crisis averted.

Or is it? Can we use this breakdown of simultaneity to come up with an impossible scenario? We can certainly try. For example, we can modify the thought experiment slightly to try to force a contradiction. Instead of having two sensors, one at the front of the barn and one at the rear, each triggering its respective door, imagine that there is only one sensor at the front. When the sensor senses that the butt end of the spear has passed the threshold, it slams the front door and only then signals the back door to open. For a split second, both the front door and the back door *must* be shut at the same time before the back door opens. No longer are the events independent, because, in a sense, the shutting of the front door *causes* the back door to open. Swapping the order of these two events *would* be a violation of the laws of physics.

This is because *causality* must be preserved, even in the upside-down world of relativity. Imagine that an assassin shoots a general with a bullet. The bullet strikes the general and kills him; had the gun not been fired, the general would not die. But if there were a nearby, fast-moving observer whose motion saw the bullet strike before the gun was fired, he might be able to knock the gun out of the assassin's hand before the gun is fired. He might be able to prevent the assassination he just saw! It's as if he traveled back in time and changed the past. This makes no sense, even in the strange domain of modern physics.

There is a limit to the reordering of events in relativity. If an event (1) *causes* an event (2), there is no way that an observer can see (2) before he sees (1). These two events are said to be *causally connected*. Even taking into account relativity's distortion of time, a traveler moving near the speed of light will never see a reversal of causally connected events. He would never see your birth before he sees your mother's; your mother's birth must come before yours, because your mother's existence *causes* your birth. Similarly, in the modified spear-in-the-barn paradox, the front door's closure causes the back door's opening. Therefore, from any point of view—from a stationary observer's or the sprinter's—the back door must open after the front door. With this modified sensor, let's rerun the scenario.

From the sprinter's point of view, the back door only opens when the front sensor is triggered—when the butt end of his spear crosses the threshold. The front part of his spear, which is fifteen meters long, will smash through the back door before it triggers the sensor that closes the front door. The outcome is a collision, at least from the sprinter's point of view.

Aha! Now it looks like we've got Einstein in a tight spot because, as before, from the stationary observer's point of view it seems possible that the spear fits well within the barn, giving enough time to open the door and avert the collision. In one frame of reference, there's a smack-up, and in the other, nothing! That's a contradiction. Or so it would seem. There is a way out, an additional subtlety that we have to take into account. And this is where information theory begins to reveal itself.

The sensor at the front of the barn has to signal the back door to open. It has to transmit information—the command to open—from the front of the barn to the rear of the barn. At least one bit of information must travel from the front of the barn to the back, and information cannot travel from place to place instantly, because information has a physical presence. Transmitting this bit takes time. In the stationary observer's frame of reference, the front sensor slams the door shut and sends a message to the back door. However, the tip of the sprinter's pole has a nine-meter head start and is zooming toward the rear door at 80 percent of the speed of light. That's a tough headstart to overcome. In fact, unless the message travels *faster* than the speed of light, there is no way it can make up the distance fast enough. The signal to the back door gets there too late: the spear strikes the door before the message arrives. So, even from the point of view of the stationary observer there's a jarring collision. Both observers agree; a smack-up occurs. Paradox averted. Averted, that is, so long as information can travel no faster than the speed of light.

Einstein's theory holds firm—but only when there is a limit on how fast information can travel. If, somehow, information could travel faster than the speed of light, causality would break down; you would be able to send a message into the past and affect the future. So long as information behaves itself and moves at light speed or below, Einstein's theory is completely consistent.

This is what is behind the famous "nothing can go faster than the speed of light" dictum, but in fact, that dictum is an oversimplification. Some things *can* go faster than the speed of light. Even light itself can break light speed, in a sense. The true rule is that *information* can't travel faster than the speed of light. You cannot take a bit of information, transmit it, and have it get to a recipient faster than a beam of light can make the same trip, otherwise causality will break down. The ordering of events in the universe would no longer make sense; you might be able to build a time machine and be born before your mother.

The seeming paradoxes in relativity hinge upon the transfer and motion of information; relativity, deep down, is a theory about information. Sometimes its rules are incredibly subtle, but they have held, despite legions of scientists who for the past century have tried to find loopholes. The puzzle of faster-than-light travel is a puzzle of information. So is the problem of time travel.

In an unassuming laboratory in New Jersey, scientists built the first time machine. Lijun Wang, a physicist at NEC Research Institute outside Princeton, sent a pulse of light faster than light speed—and forced it to exit a chamber before it ever entered.

This is no joke. It was published in the peer-reviewed journal *Nature* in 2000 and has been replicated by a handful of labs across the country. It is not that difficult an experiment to perform: all it requires is a chamber full of gas, a laser, and a very precise stopwatch. And while Wang's work is the most dramatic example of breaking the speed of light, it is not the only one. Barely a month before Wang's experiment, Italian physicists used a clever geometric construction to get a laser beam to exceed *c*, the speed of light. Half a decade before that, Raymond Chiao, a physicist at the University of California at Berkeley, used a bizarre quantum-mechanical property called *tunneling* to make a light pulse go faster than *c*.

The easiest faster-than-light experiment to understand is one that was performed in Italy in 2000. In it, Anedio Ranfagni and colleagues at the Italian National Research Council in Florence took a beam of microwaves, passed it through a ring, and then bounced it off a curved mirror to create what is called a *Bessel beam* of microwave light. Viewed from above, a Bessel beam has planes of waves that intersect like an X. The scientists watched as the intersection of that X moved more than 7 percent faster than the speed of light; it looked as if they were sending something—the intersection—faster than *c*. (An easy way to see what's going on is to make an X with your two index fingers nearly parallel to each other. Move your hands apart slowly and you'll see that the intersection moves up your fingers at a speed that's much greater than the speed at which your hands are moving apart.) But what happens if you try to send a message with this scheme? Can it go faster than light?

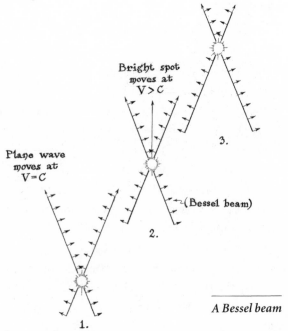

A Bessel beam

Einstein would be relieved: the answer is no. Imagine, for example, that Alice is a sentinel on Jupiter. When she spots an invading force of bug-eyed creatures from Alpha Centauri, she needs to send word back to Earth. Luckily for her, a Bessel beam communication line is already set up between Jupiter and Earth; she needs to use it to send back a single bit of information, a warning. A sudden change in signal—the beam blinking all of a sudden, for example—would suffice; when the beam flickers once, it signals the transmission of one bit of information. It means that the alien redcoats are coming.

Alice can disrupt the beam by sticking her hand in the center of the beam, absorbing the light and making the point of intersection go dark. Since the intersection moves faster than the speed of light, shouldn't that dark spot zoom down the beam faster than the speed of light, too? Well, no. Because of the way the beam is set up—the plane waves move at an angle, even though the intersection moves straight at Earth—the dark spot moves away from the center of the beam, and the intersection itself remains bright. Someone monitoring the beam on Earth would not see the center of the beam flicker at all; no matter how Alice fiddles with the intersection of the beam on Jupiter, the beam will never flicker on Earth. The bit is lost, sent out into space at the speed of light, and nobody on Earth will ever receive Alice's message. Though the intersection moves faster than the speed of light, it cannot carry a bit. It carries no information.

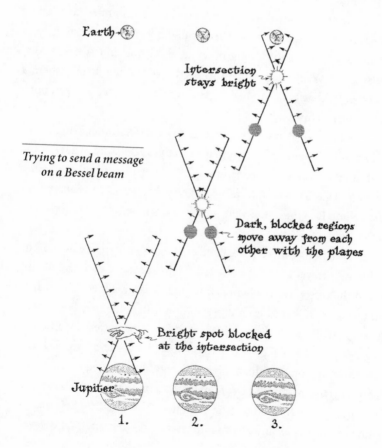

Trying to send a message on a Bessel beam

There is another option for Alice, though. She can block the two spots on the beam that wind up intersecting on Earth. In that case, as the light waves move, the two spots move closer and closer together as they travel toward Earth, eventually converging at the receiver, which suddenly sees the beam turn off. In this case, Earth would receive one bit of information—but that information only moved as fast as the speed of light. Remember that the plane waves themselves only move as fast as *c;* the intersection is the only "object" that moves faster than the speed of light. The blocked spots on the beam would travel at the speed of light toward Earth. Alice's message travels at the speed of light—no faster.

Thus, the Italian Bessel beam experiment amounted to nothing more than a geometric trick. There really isn't anything moving faster than the speed of light. On the other hand, it's not quite so easy to dismiss Lijun Wang's experiment. Indeed, his baffling setup, where a pulse of light exits a chamber of gas before it enters, seems like a bona fide time machine.

The heart of Wang's time machine is a six-centimeter-long container full of cesium gas. Cesium is a reactive metal, somewhat like the sodium that is used in streetlamps. When set up properly, the chamber of cesium gas has a very peculiar property that is known as *anomalous dispersion.* It is this effect that turns the chamber into a faster-than-light device.

In a vacuum, different frequencies of light—different colors of the rainbow—travel at the same speed. Light speed, of course. But this is not the case when the light is traveling through a medium, such as air or water. In that case, light moves slower than the speed of light in a vacuum. In most cases, different colors get slowed to different degrees. Redder light—lower-frequency light—tends to feel the effects of matter less than higher-energy, higher-frequency, bluer light. This means that reddish light tends to move through a box of air or a glass of water slightly faster than does bluish light. This effect is known as *dispersion* and it is an important effect because of the wavelike part of light's nature. (It's also the cause of the separation of colors in a rainbow.)

Any wavelike object, such as a pulse of light, can be dissected into its component parts—in light's case, into beams of light of different frequencies.[6] In most regions of space, these beams cancel out each other, but in one region the beams of different frequencies reinforce each other, creating a pulse of light. The pulse moves because the individual beams of light are moving at c; their regions of reinforcement and cancellation shift forward at the speed of light, causing the pulse to move forward at the speed of light, too. At least that is what happens in a vacuum. It's a tad more complicated in a dispersive medium, like air.

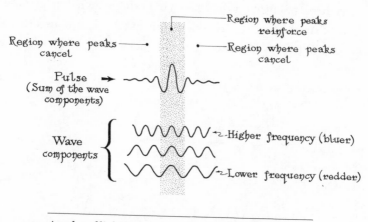

A pulse of light in a vacuum and its Fourier transform

Since air slows down reddish light somewhat less than bluish light, it messes up the cancellation somewhat. As a pulse gets very close to a chamber full of air, the cancellation stops being perfect: the pulse spreads out a little as the high-frequency waves slow down relative to

6. This technique is known as a Fourier transform, after its inventor, Jean-Baptiste-Joseph Fourier. Fourier was nearly sent to the guillotine in 1794, during France's Reign of Terror, and eventually wound up as a science adviser to Napoleon.

the low-frequency ones. The pulse in the chamber gets fatter and fatter as it moves, and then it emerges from the far side of the chamber much fatter than when it entered. The pulse has become distorted by the medium.

However, the story gets really weird if the medium is not dispersive in the ordinary way, slowing blue light more than it slows red light. If instead of ordinary dispersion it has anomalous dispersion, where the opposite happens, red light is slowed more than blue light, and a time machine is the result.

As before, when a pulse approaches the chamber of cesium gas, the medium messes up the nice cancellation. In the previous air example, where the blue light traveled slower than the red light, the cancellation got destroyed near the pulse, causing the pulse just to spread out. In the cesium chamber, on the other hand, where the red light travels slower than the blue, the cancellation is messed up a long distance away from the pulse. It is as if the pulse suddenly appears a long way

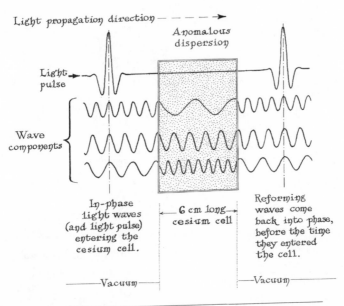

*A pulse of light in an anomalous dispersion medium
and its Fourier transform*

away: it's as if it moves faster than the speed of light. In fact, if the anomalous dispersion effect is sufficiently pronounced, the pulse can emerge from the chamber before it ever enters!

This is hard to visualize, but it is a consequence of the properties of light. There is no trick; the pulse emerges from the chamber before it enters because it moves faster than the speed of light through the chamber. In Wang's original experiment, the pulse moved at 300 times the speed of light and left the cesium gas about 62 nanoseconds before it entered. Wang's experiment and variations thereof have since been reproduced many times. There is little controversy; most physicists agree that the pulse effectively travels faster than the speed of light. For example, Daniel Gauthier, a physicist at Duke University, and two colleagues repeated the experiment with potassium vapor in the chamber instead of cesium, and sure enough a pulse exited the chamber 27 nanoseconds faster than it would if it were traveling at the speed of light; it broke the universal speed limit, c, by more than 2 percent.

Gauthier and his colleagues weren't satisfied with making pulses move faster than light; they tried to send information on those pulses. And to send information, you need to send a bit. Gauthier's team rigged the laser beam to get brighter after a short interval (encoding a **1**) or to get dimmer after that same interval (encoding a **0**). Then, on the other end of the chamber they had a detector that recorded the moment it was able to distinguish a **1** pulse from a **0** pulse with a given amount of confidence. If the chamber truly made information travel faster than the speed of light, then the detector should be able to register a **1** or **0** pulse faster than it could if the pulse merely traveled at the speed of light; the faster-than-light pulse would deposit its information at the detector faster than would a light-speed pulse.

What they found was just the opposite. Though the light-speed pulse emerged from the chamber later than the faster-than-light pulse, it deposited its information before the faster-than-light pulse. Even though the sped-up beam reached the detector first, its information lagged a little. Einstein, once more, could rest easy. Like the Bessel beam device, the gas-chamber time machine cannot transmit *information* faster than the speed of light. However, the reason is a touch more subtle than it was for the Bessel beam device. It has to do with the shape of the pulse.

When a pulse passes through a dispersive medium like air or cesium or potassium, it gets somewhat distorted. Sometimes it gets fatter; sometimes it gets thinner. It gets taller in places, and shorter in others. In Gauthier's experiment, the **0** pulse and the **1** pulse got distorted in slightly different ways. The original **0** pulse was contrived to drop off suddenly, but after the distortion caused by its passage through the chamber, it dropped off somewhat less abruptly. The **1** pulse, on the other hand, got brighter less abruptly after going through the chamber of cesium. A slower-dimming **0** and a slower-brightening **1** meant that the two possibilities could not be distinguished as quickly; it was harder to tell the difference between the **0** pulse and the **1** pulse. Even though the pulses moved through the chamber faster than light speed, it took longer for the detector to tell them apart because of that distortion—more than compensating for the speedup of that pulse. In sum, the place where the information in the pulse resides—the place where the bit sits in the pulse—always moves slower than the speed of light, even when the pulse itself breaks the speed limit.

This same effect derails attempts to transmit information with yet another faster-than-light technique that exploits a bizarre feature of the subatomic world known as quantum tunneling. In the classical world, throw a ball at a concrete wall and it will bounce right off. In the quantum world, throw a particle such as a photon at an impenetrable barrier and it will bounce right off. Most of the time. Once in a long while—how probable this is depends on the nature of the barrier and the particle—the particle will pass right through the barrier. A photon, for example, can "tunnel" right through a wall of a sealed box, even though the classical rules of physics would forbid any light from getting in or out at all. This is a consequence of the mathematics of quantum mechanics, and it has been observed numerous times. In fact, the process of radioactive decay is a form of tunneling; for example, an alpha particle (a conglomerate of two protons and two neutrons) is effectively trapped within the box of an unstable atomic nucleus such as uranium-238. The alpha particle rattles around inside that nucleus for years and years (on average, about four and a half billion years), and then, all of a sudden, it pops right through the side of the box and flies away. The escaping alpha particle then makes a click when it slams into a sensitive radiation detector.

What makes tunneling interesting to the faster-than-light crowd is that the process occurs incredibly fast, perhaps even instantly. It can suddenly tunnel through a barrier—snap!—without actually taking the time to pass through the barrier.[7] It's as if the particle disappears and appears somewhere else, all at the same moment.

Raymond Chiao, the University of California physicist, set up an experiment where he sent photons at a relatively thick barrier, a coated slice of silicon. Once in a while, a photon tunneled through that barrier. Sure enough, Chiao measured the speed that the photons traveled through that barrier and discovered that they were, indeed, moving faster than the speed of light. However, Chiao realized that those photons could not be used to transmit information faster than light speed. Just as in the gas-chamber device, the photon's "shape" changes as it passes through the barrier. A photon is no different than a pulse in the quantum-mechanical view of the world; it behaves like a packet of waves as much as it does a particle. And that wave packet gets shifted and reshaped as it tunnels through the barrier. In fact, only the leading edge of the photon wave packet gets through the barrier; on the far side, the wave packet is much smaller, and the leading edge has been reshaped so that any bit that you encode on that pulse of light is knocked backward by the reshaping process. Einstein, once again, would have been relieved. Information does not travel faster than the

7. Yes, this seems incredible, but it is a consequence of the laws of quantum mechanics, which will be explained in more detail in the next chapter.

speed of light, even though the photon itself does. The theory of relativity had withstood all attempts to transmit information faster than light speed.

However, there is another threat to relativity posed by the laws of quantum mechanics. It was a threat that Einstein himself discovered, and one that caused him to reject the theory he helped create. Here, too, the key to understanding is information.

CHAPTER 6

PARADOX

Natura non facit saltus. (Nature does not make jumps.)

—Gottfried Wilhelm Leibniz

Relativity is a theory that deals with information. Einstein's equations dictate a speed limit for the transfer of information across space: the speed of light. They also explain why different observers will gather seemingly contradictory answers when asking the same questions—even when all are collecting information about the same events. Relativity changed scientists' way of looking at the universe, at how objects interact with one another over great distances, at high speeds, and under conditions of very strong gravity. It was Einstein's greatest triumph. But it would not land him the Nobel Prize.

Einstein's Nobel came *despite* his work on relativity theory, which, at the same time that it was embraced by many of the greatest thinkers of the day, was rejected by some of the more rigid and conservative members on the prize committee. Nevertheless, Einstein did win a Nobel in 1921 for one of his other great insights: the quantum theory of light.

Ironically, Nobel Prize notwithstanding, Einstein came to despise the quantum theory that he helped create. With good reason. The faster-than-light challenges to relativity theory are child's play compared to those that come directly out of the laws of quantum theory. Einstein himself found a major one of these challenges—a quantum-theoretical trick that seemed to tear a hole in his lovely theory of relativity. To his horror, he spotted what seemed like a logical loophole in quantum mechanics; at first glance, it appeared that physicists could exploit that loophole to send information faster than the speed of light. If true, it would allow engineers to build the equivalent of a time machine. The laws of quantum theory, it seemed, would give scientists the ability to alter the past and change the future.

Einstein thought that this loophole, this mysterious cosmic messaging system that he discovered, would prove that the theory of the quantum world was absurd and would have to be discarded. He was wrong. Countless times, physicists have seen Einstein's mysterious quantum "spooky action" link two particles. If such particles can somehow communicate, they must do so at several thousand times the speed of light. Einstein's nightmare is real.

The strange tricks that Nature plays with quantum objects have been observed, and quantum theory's weirdest predictions have been verified. The informational paradoxes of quantum mechanics drove Einstein away from quantum theory, and he would never see them resolved. They are still the most troubling features of quantum theory, and only now, nearly a century later, are scientists beginning to understand them, thanks to the science of information.

Though Einstein loved relativity and hated quantum theory, both are his children. The siblings came from the same source: they are each tied to thermodynamics and information, and they were both born in light.

Like relativity, the story of quantum theory goes back to Young's experiment in 1801, which seemed to end the debate about whether light is a particle or a wave. Young showed that light makes an interference pattern when a beam passes through two slits simultaneously. This is what waves do. It is not what particles do.

With Young's experiment, you can make the beam dimmer and

dimmer, yet no matter how dim the beam of lights gets, the interference pattern remains. If light were, in fact, composed of particles, at some point—when the beam is sufficiently dim—only one particle of light would be going through the slits at a time. Yet if you were to perform the experiment over and over again, you would never see that single electron hit certain places on the screen: there is always an interference pattern. Even with a single particle going through the slits at a time, it somehow interferes with itself, preventing itself from striking certain regions of the detector.

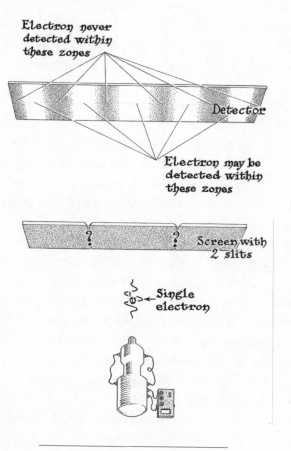

Electron never detected within these zones

Detector

Electron may be detected within these zones

? ? Screen with 2 slits

Single electron

A single electron interfering with itself

How could a single particle cause an interference pattern? How could an indivisible corpuscle possibly interfere with itself? Common sense says that it can't. If light were a particle, the interference pattern should suddenly disappear when the beam gets too dim. This is not what happens. The interference pattern remains; therefore, scientists concluded, light must be a wave and not a particle. Maxwell's equations, which were very similar to the equations that showed how a water wave propagates in the ocean, reinforced the idea. Light behaves like a wave; it is described by wavelike mathematics; therefore, it must be a wave and not a particle. Case closed.

Well, not *entirely* closed. There were a few problems with assuming that light was a wave. The most important of these problems emerged in 1887, when the German physicist Heinrich Hertz discovered a curious phenomenon: when he put a metal plate in a beam of ultraviolet light, it sparked. The light tore electrons right out of the metal. This *photoelectric effect* could not be explained by the wave theory of light just as surely as the interference fringes could not be explained by the particle theory of light. Wave theory's failure to explain Hertz's sparking has to do with energy. The sparks in the metal were caused by light knocking electrons away from metal atoms—by the energy that the light contained.

On its face, the problem seems far removed from the ideas of information and thermodynamics, but as with relativity, information theory will come rushing in once the problem is fully understood. Indeed, the explanation for why metals spark would lead to the biggest philosophical problem in physics today, a problem that has to do with the roles of measurement and of information in the way the universe works.

However, to a late-nineteenth-century physicist, figuring out why a metal sparks seemed not to have any great significance. In fact, it looked like a problem no different from figuring out why a ball falls back to the ground after it is batted into the air. An electron is bound to its atom by a certain amount of energy, just as a baseball is bound to the Earth by gravity. If you are going to free an electron from an atom, you have to supply the energy to break that bond, just as you have to give a baseball a big enough whack to send it out of the atmosphere. If a subatomic whack doesn't have enough energy, an electron flies away from its atom a bit but sinks right back down again, just as a baseball propelled upward with insufficient force must come back to Earth. However, if you give the electron more punch than the binding energy of the atom, the force sends it out of the atom entirely (the baseball zooms into orbit).

In the photoelectric effect, the source of the electron-whacking energy has to come from light. Now, for the moment, assume light is a wave. If so, buffeting light waves must deposit their energy in the electrons, giving them a bunch of energy. This energy induces the electrons to leap away from the metal atoms. If the waves don't deposit enough energy into the electron, if the waves' collective energy is below the required threshold, then the electrons will stay put. However, if the waves are sufficiently energetic, then they will induce the metal to spark. So far so good.

In wave theory, though, there are two ways to increase the energy of an incoming bunch of waves. The first way is pretty easy to see: just make the waves bigger. One-foot ocean waves pack less punch than three-foot waves, and ten-foot waves can knock a swimmer senseless. A wave's height is known as its *amplitude;* the bigger the wave, the larger the amplitude and the more energy it carries. With water waves, amplitude translates to physical height, but in other sorts of waves it might have a different interpretation. With sound waves, for example, amplitude is related to volume: the louder a sound, the greater the amplitude of the sound waves. And with light, amplitude is related to brightness. A bright yellow beam has a greater amplitude than a dim yellow beam.

The second way to increase energy in a set of waves is a little more subtle: make the waves more frequent. If the wave crests are closer together—if more waves are striking the beach per minute—the waves transfer more energy to the shore. So, the greater the frequency of the waves, the more energy they contain. With light, frequency corresponds to color. The lower-frequency light—infrared, red, and orange light—contains less energy than yellow, green, blue, or violet light, which have greater frequencies. Ultraviolet light and X-rays have more energy still, as their frequencies are even greater than those of visible light.

So, if there is an energy threshold to knock electrons out of a metal, there should be two ways to get light above that critical threshold. With a beam of given brightness, you can change the frequency of light from red to green to blue to ultraviolet, and at some point the electrons should start leaping out of the metal. Sure enough, this is what happens. Red light didn't cause sparking in Hertz's metal sheet, nor did green or blue. But when the light color became a high enough frequency—when the beam was ultraviolet—the sparking suddenly began.

The second way to get the light above that energy threshold is to fix the frequency of the beam—keeping it, say, at the same shade of yellow—but increasing the brightness of the beam. If you start off with a dim beam of yellow, it won't have enough energy to induce the electrons to leap out of the metal. But as you brighten the beam, it gets more and more energetic. When the beam finally gets bright enough, when the amplitude of the beam gets big enough, the electrons should suddenly start getting knocked loose and the sparking should begin. This is *not* what happened.

No matter how bright a yellow beam was, it never freed electrons from the metal. Worse yet, even the dimmest ultraviolet beam—which, according to the wave theory of light, shouldn't have enough energy in the metal to free electrons—caused sparking. Just as it made no sense for a single particle of light to make an interference pattern, it made no sense for a dim wave of ultraviolet light to be able to knock electrons loose while a bright yellow beam could not. In wave theory, there should be an amplitude threshold for the photoelectric effect just as there was a frequency threshold. But Hertz's experiment showed that only frequency seemed to matter. This contradicted the wave equations of light that scientists had long since accepted.

Physicists were stuck. They couldn't explain interference with the particle theory of light, and they couldn't explain the photoelectric effect with the wave theory. It took nearly twenty years to figure out what was wrong, and when Einstein did—the same year, 1905, that he formulated the special theory of relativity—he destroyed the wave theory of light forever. In its place was a new theory, quantum theory. It was Einstein's explanation of the photoelectric effect that earned him his Nobel Prize and placed quantum theory firmly in the mainstream of physics.

Einstein's work breathed life into a rogue idea that had been born five years earlier when Max Planck, a German physicist, came up with a method for resolving a mathematical dilemma. This dilemma, too, had to do with the behavior of light and matter. The equations that described how much radiation a hot chunk of matter emits—the ones that describe why a blacksmith's iron glows red and the filament in a lightbulb glows white—were not working. These equations broke down under certain conditions, shredding the theory in a cloud of mathematical infinities. Planck came up with a solution, but it came at a price.

Planck made an assumption that seemed physically absurd. He assumed that under certain circumstances matter could only move in certain ways: it is quantized. (Planck coined the term *quantum,* after the Latin word for "how much.") For example, the quantization of an electron's energy around an atom meant that an electron could only take on some energies and not others. This sort of thing doesn't happen in everyday life. Imagine what would happen if your car's speeds were quantized: if it could go 20 and 25 miles an hour, but couldn't drive at 21 or 23 or any other speed in between. If you were driving at 20 miles an hour and you pressed down on the accelerator, absolutely nothing would happen for a while. You would keep tooling along at 20 miles an hour ... 20 miles an hour ... 20 miles an hour ... and then, suddenly, pop! You would instantly be driving at 25 miles an hour.

Your car would have skipped all the speeds in between 20 and 25. This obviously doesn't happen. Our world is smooth and continuous, not jerky and jumpy. Planck himself called his quantum hypothesis an "act of desperation." However, as strange as this idea—the quantum hypothesis—was, it banished the infinities that plagued the radiation equations.

Einstein solved the photoelectric puzzle by applying the quantum hypothesis to light. Contrary to what nearly all the physicists of the previous hundred years had assumed, Einstein postulated that light is not a smooth, continuous wave but chunky, discrete particles now known as photons. This was despite the evidence to the contrary, including Young's interference experiment. In Einstein's model, each particle carries a certain amount of energy proportional to its frequency; double the frequency of a photon and you double the energy it carries.[1] Once you accept that idea, you can do a great job of explaining the photoelectric effect.

In Einstein's picture, each photon striking the metal can give the electron a kick, and the more energy the photon has, the bigger the kick. As before, the energy must meet a threshold. If the energy of the photon is too small, below the binding energy of the electron, the electron cannot escape. If the energy is large enough, though, the electron escapes. As did the wave theory of light, Einstein's hypothesis explains the wavelength threshold: if the photons do not have enough energy, then they cannot knock the electrons away from the atoms. But unlike the wave theory, Einstein's quantum theory of light also explains the *lack* of an amplitude threshold. It explains why merely increasing the brightness of the beam cannot make electrons start escaping the metal.

If the beam is made up of individual particles of light, increasing the brightness just means that more of these particles are in the beam.

---

1. Einstein came to this conclusion by trying to figure out the entropy of the light that streams off a theoretical object known as a blackbody. The roots of quantum theory are tied very tightly to thermodynamics and statistical mechanics.

Only one photon is likely to strike an atom at a time, and if that photon doesn't have the required energy, it can't knock the electron away—no matter how many photons there are in the surrounding neighborhood. It is one photon per atom, and if the incoming photon is too weak, nothing happens, regardless of the brightness of the beam.

Einstein's quantum theory of light explained the photoelectric effect in wonderful detail; the hypothesis completely explained the puzzling experimental observations that could not be explained by the wave theory of light.[2] This was really puzzling to physicists at the time: Young showed that light behaves like a wave and not a particle, but Einstein showed that light behaves like a particle and not a wave. The two theories were in direct conflict, and they couldn't both be right. Or could they?

Just as with relativity theory, information was at the heart of the problem. In relativity theory, two different observers can gather information about the same event and get mutually contradictory answers. One might say that a spear was nine meters long while the other might say that it was fifteen meters long, and both can be right. In quantum theory, there is a similar problem. An observer, measuring a system in two different ways, might get two different answers. Do an experiment in one way and you might prove that light is a wave and not a particle. Do a similar experiment in a slightly different way and you can prove that light is a particle and not a wave. Which is right? Both—and neither. The way you gather the information affects the outcome of the experiment.

Quantum theory can be cast in the language of information theory—

2. It explained other effects, too, such as the so-called Stokes's rule for fluorescent materials. If you zap certain minerals, such as some forms of calcite, with high-energy light, they will glow. Stokes's rule says that the glow is always redder—a lower frequency—than the light that you shine on the mineral. This is hard to explain with the wave theory of light, but it's easy to explain with quantum theory: when a particle of light deposits its energy in an atom, the atom reemits that energy. The packet of energy that it emits must be less than or equal to the energy it absorbs; the frequency of the emitted photon must be less than or equal to the frequency of the absorbed photon.

in talk about transferring information (including the 1s and 0s of binary choices)—and when it is, it reveals a whole new depth to the paradoxes of the quantum world. The conflict between waves and particles is just the beginning.

Einstein's theory brought Planck's quantum hypothesis into the mainstream, and over the next three decades Europe's best physicists developed a theory that did a beautiful job of explaining the behavior of the subatomic world. Werner Heisenberg, Erwin Schrödinger, Niels Bohr, Max Born, Paul Dirac, Albert Einstein, and others built up a set of equations that explained with stunning precision the behavior of light and electrons and atoms and other very tiny objects.[3] Unfortunately, though this framework of equations—quantum theory—always seemed to get the right answers, other consequences of those equations seemed to contradict common sense.

The dictates of quantum theory, at first glance, are ridiculous. The strange, seemingly contradictory properties of light are par for the course. Indeed, they come directly from the mathematics of quantum theory. Light behaves like a particle under some conditions and like a wave under other conditions; it has some of the properties of each, yet is neither truly particle nor wave.

It is not only light that behaves this way. In 1924, the French physicist Louis de Broglie suggested that subatomic matter—particles like electrons—should have wavelike properties as well. To experimentalists, electrons were *obviously* particles, not waves; any half-competent observer could see electrons leave little vapor trails as they streaked from one end of a cloud chamber to the other. These trails were clearly the tracks of little chunks of matter: particles, not waves. But quantum theory trumps common sense.

---

3. Sometimes, the precision is utterly astounding. For example, theory predicts how an electron twists in a magnetic field. Plug in the numbers and you will discover that the theory matches observation to nine decimal places. It is as if theory predicted the distance between the Earth and the moon with an uncertainty of about a meter.

Though the effect is much harder to spot with electrons than it is with light, electrons *do* show wavelike behavior as well as their more familiar particle-like behavior. In 1927, English physicists shot a beam of electrons at a crystal of nickel. As electrons bounce off regularly spaced atoms and zoom through the holes in an atomic lattice, they behave as if they have just passed through the slits of Young's experiment. Electrons *do* interfere with each other, making an interference pattern. Even if you ensure that only a single electron at a time strikes the lattice, the interference pattern persists; the pattern cannot be caused by electrons bouncing off each other. This behavior is *not* consistent with what you would expect of particles: the interference pattern is an unmistakable sign of a smooth, continuous wave, rather than discrete, solid particles. Somehow, electrons, like light, have *both* wavelike and particle-like behavior, even though the properties of waves and of particles are mutually contradictory.

This twofold wave-particle nature is true of atoms and even molecules just as it is true of electrons and light. Quantum objects can behave like waves as well as particles; they have wavelike properties and particle-like properties. At the same time, they have properties *inconsistent* with being a wave and with being a particle. An electron, a photon, and an atom are both particle and wave, and neither particle nor wave. If you set up an experiment to determine whether a quantum object is a particle, **1**, or a wave, **0**, you will get a **1** sometimes or a **0** sometimes, depending on the experiment's setup. The information you receive depends on how you gather that information. This is an unavoidable consequence of the mathematics of quantum mechanics. It is known as wave-particle duality.

Wave-particle duality has some really bizarre consequences—you can use it to do things that are absolutely forbidden by the classical laws of physics—and this seemingly impossible behavior is encoded in the mathematics of quantum mechanics. For example, the wavelike nature of the electron allows you to build an interferometer from electrons just as you can build one from light. The setup is pretty much the same in both cases. In the matter-wave interferometer, a beam of particles, such as electrons, shoots toward a beam splitter and goes off in two directions at once. When the beams recombine, they either reinforce each other or cancel out each other, depending on the relative sizes of the two paths. If you tune the interferometer properly, you should never, ever spot an electron at the detector, because the beams moving down the two paths can completely cancel each other. This cancellation works no matter how dim the electron beam is, no matter how few electrons are striking the beam splitter. In fact, if you set up your apparatus correctly, even if a single electron enters the interferometer and hits the beam splitter you will never detect the electron emerging from the other side.

Common sense would tell you that an indivisible particle like an electron would have to make a choice at the beam splitter: it would have to choose to take path A or path B, to go to the left or to go to the right, but not both. It should be a purely binary decision; you can assign a **0** to path A and a **1** to path B. The electron would travel down its chosen path. It would make its choice, **0** or **1**, and then, at the far end of the interferometer, it would strike the detector. Because only a single electron travels through the interferometer, there should be nothing to interfere with it, no other particles to block it. Regardless of whether the electron chooses path A or path B, it should emerge on the other side at the detector unhindered. The interference pattern should disappear. There are no other particles that can interfere with the electron. But that is not what happens; common sense fails.

Even when a single electron at a time enters the interferometer, there's an interference pattern. Somehow, something is blocking the electron. Something prevents the electron from emerging from the beam splitter in certain ways and striking the detector in some places, but what could that something be? After all, the electron is the only thing in the interferometer.

The answer to this seeming paradox is hard to accept, and you will have to suspend your disbelief for a moment, as it sounds impossible. The laws of quantum mechanics reveal the culprit. The object that blocks the electron's motion is the electron itself. When the electron hits the beam splitter, it takes both paths at once. It doesn't choose to take path A or path B; instead, it goes down both paths simultaneously, even though the electron itself is indivisible. It goes left and right at the same time; its choice is simultaneously a **0** and a **1**. Faced with two mutually exclusive choices, the electron chooses both.

In quantum mechanics, this is a principle known as *superposition*. A quantum object like a photon or an electron or an atom can do two (classically) contradictory things or, more precisely, be in two mutually exclusive quantum *states,* simultaneously. An electron can be in two places at once, taking a left path and a right path at the same time. A photon can be polarized vertically and horizontally at the same time. An atom can be both right side up and upside down (more technically, its *spin* can be up and down) at the same moment. And in information-theoretic terms, a single quantum object can be a **0** and a **1** simultaneously.[4]

This superposition effect has been observed many times. In 1996, a team of physicists at the National Institute of Standards and Technology laboratories in Boulder, Colorado, led by Chris Monroe and David Wineland, made a single beryllium atom sit in two different places at the same time. First, they set up a clever laser system that separated objects with different spins. When the lasers struck an atom with an upward spin, they pushed it a tiny bit in one direction, say, to the left; when they struck an atom with a downward spin, they pushed it in the opposite direction, a hair to the right. The physicists then took a single atom, isolated it carefully from its surroundings, and bombarded it with radio waves and lasers, putting it in a state of superpo-

---

4. This is *not* the same as being halfway between **0** and **1**—say, 0.5. This is easy to see if you think of it in terms of direction. If **0** is to the left and **1** is to the right, 0.5 would be straight ahead. But a superposition of **0** and **1** is left *and* right at the same time, something that is impossible for a classical, indivisible object like a person.

sition. It was in both a spin-up state and spin-down state, both **1** and **0**, at the same time. Then they turned on the laser separating system. Sure enough, the same atom, spin up and spin down at the same time, moved to the left *and* to the right simultaneously! The atom's spin-up state moved to the left, and its spin-down state moved to the right: a single beryllium atom was in two places at once. A classical, indivisible atom could never be *both* **1** and **0** at the same time, but the Colorado team's data indicated that the atom was simultaneously in two positions, fully 80 nanometers—about ten atomic widths—away from each other. The atom was in a (dramatic) state of superposition.[5]

Superposition explains how a single electron can make an interference pattern even though a single classical object never could. The electron interferes with itself. When the electron hits the beam splitter, it enters a state of superposition; it takes path A and path B at the same time, it chooses both **0** and **1**. It is as if two ghostly electrons travel down the two sides of the interferometer, one on the left and one on the right. When the two paths rejoin each other, the ghostly electrons interfere with each other, canceling out each other. The electron enters the beam splitter but never emerges, never strikes the detector, because the electron takes two paths simultaneously and cancels itself out.

If this effect weren't strange enough, it gets even more bizarre. Superposition is fragile and slippery. As soon as you peek at a superposed object, as soon as you try to get information about whether, say, an electron actually is a **0** or a **1**, is spin up or spin down, or takes path A or path B, the electron suddenly and (seemingly) randomly "chooses" one path or the other. The superposition is destroyed. For example, if you rig the two paths of an interferometer with a trip

5. There are a number of different interpretations of quantum theory; physicists disagree about what it *really* means for a quantum object to be in two places at the same time. (In this book, I have chosen an interpretation that I think makes the text clearest—more on this in chapter 9.) Regardless, all interpretations agree that you cannot explain quantum behavior by cramming it into a classical framework. Quantum theory really does force you to discard the commonsense ideas of classical physics in some manner. All interpretations have quantum objects in superposition; they just have different notions of what that superposition represents.

wire—say, a laser beam that illuminates path B and sends a bit **1** to a computer when the electron crosses the beam—the electron cannot be in superposition. It "chooses" to go down path A or path B rather than both, "chooses" to be a **0** or **1** rather than both, and the interference pattern disappears. Without that trip wire, the electron will be in a state of superposition, taking two paths at once. But the instant you extract *information* about the electron's path, attempting to detect it or measure it, the superposition evaporates—the superposition *collapses.*[6] As soon as information about the electron's path leaves the interferometer system, the electron instantly and randomly chooses the left path or the right path, **0** or **1**, as if God had flipped a cosmic coin to settle the matter.

The principle of superposition is so strange that lots of physicists had trouble accepting it, even though it explained observations that could not be explained in any other way. How could a single electron take two paths simultaneously? How can a photon be spin up and spin down at the same time? How can an object take two mutually contradictory choices? The answer—which was not yet known—had to do with information; the act of gathering and transmitting information is where scientists found the key to understanding the unsettling and counterintuitive idea of superposition. However, in the 1920s and 1930s scientists were not yet armed with the formal mathematics of information theory, yet they were not completely helpless. When confronted with the paradoxical idea of superposition, physicists whipped out their favorite weapon—the thought experiment—to try to destroy the concept. The most famous of these came not from Einstein, but from the Austrian physicist Erwin Schrödinger.

6. Interestingly, this works even if you rig only one of the two paths, say, B, with a trip wire. If you send an electron into the interferometer and it chooses path B, the laser detects the passage of the electron, and you get one bit of information about which path it took. If it chooses path A, you don't have any equipment set up to detect its passage, but the nonclick of the detector tells you that it didn't take path B: it took path A. So, even though nothing passed by the laser, you still get one bit of information. The trip wire on B destroyed the superposition of the electron even though the laser never touched the electron; after all, it took path A and not B.

The modern form of quantum mechanics began to take shape in 1925, when the German physicist Werner Heisenberg came up with a theoretical framework based upon (at the time) relatively unfamiliar mathematical objects known as matrices. Matrices have a property that is a little unsettling at first: they don't *commute* when you multiply them together.

When you multiply two numbers, it doesn't matter in what order you multiply them: 5 times 8 is the same thing as 8 times 5. In other words, numbers commute under multiplication. But if you multiply matrix A by matrix B, the result is often quite different from the product of matrix B times matrix A. Nowadays, physicists are comfortable with the idea of noncommutative mathematics, but at the time, Heisenberg's matrix mechanics caused a bit of a stir. This was, in part, because the noncommutative property of matrices led to a very, very strange consequence: Heisenberg's uncertainty principle.

In Heisenberg's theory, a matrix might represent a property of a particle that you can measure: its position, its energy, its momentum,[7] its polarization, or some other *observable*. In Heisenberg's mathematical framework, something odd happens if two of those matrices don't commute with each other: their information is linked in a very disturbing manner.

Position and momentum are two such observables whose matrices don't commute. In physics-speak, a particle's position and momentum are *complementary*. The mathematics of Heisenberg's theory implied that gathering information about one of a pair of complementary observables would cause you to lose information about the other. So, measure a particle's position—get information about where

7. Momentum is the measure of how much "oomph" an object has, and it is related to the object's mass and velocity. A car moving at 5 miles an hour has less momentum than does one that is moving 30 miles an hour; the 30-mile-an-hour car will give you much more of a knock if it hits you. Likewise, a truck moving 30 miles an hour has more momentum than a car moving at the same speed.

it is—and you automatically lose information about its momentum. Conversely, if you get information about a particle's momentum—if you reduce your uncertainty about how much momentum it has— you increase your uncertainty about where it is. At its logical extreme, if you somehow were able to determine with 100 percent accuracy how much momentum a particle carries, you would know *nothing* about where it is. It could be anywhere in the universe. This is the famous uncertainty principle.

To classical physicists, this was a very unappealing idea. It meant that it is utterly impossible to have perfect information about two complementary observables at the same time. You cannot know an atom's position and momentum simultaneously; you can have perfect information about one, but that means you have no information about the other. It's an inherent limit to human knowledge.[8] And scientists hate limits.

Even though Heisenberg's mathematical framework explained, beautifully, the strange world of the very small—the world of quantum objects—the matrix theory had a lot of violations of common sense. Heisenberg's uncertainty principle was bizarre, and superposition was utterly appalling. It is no surprise that Heisenberg's quantum theory had a lot of enemies. Chief among them was Schrödinger.

Schrödinger so hated Heisenberg's matrix mechanics that he decided to take a vacation—with his mistress. He took her up to a chalet in the Swiss Alps and came down the mountain armed with an alternative to Heisenberg's matrix theory.[9] Unlike Heisenberg's framework, Schrödinger's version of quantum theory was based upon mathematical objects that physicists were familiar with: integral and differential equations just like those of Newtonian mechanics and Maxwell's equations. Instead of describing quantum objects in terms of matri-

---

8. In fact, it limits Nature as well . . . as we shall see in the next chapter.
9. The mathematician Hermann Weyl famously described Schrödinger's discovery of the theory as a "late erotic outburst."

ces, Schrödinger's method used a mathematical construct that behaved like a wave. This construct, a *wave function*, described all of an object's quantum-mechanical properties without resorting to mathematical exotica such as matrices. But recasting quantum theory in more familiar terms did not get rid of the strangeness of the uncertainty principle and of superposition. A few years after Schrödinger proposed his alternative theory, physicists proved that it was mathematically equivalent to Heisenberg's; even though the two theories used different types of mathematical objects, they were no different underneath all the formalism. So all the weirdness with uncertainty and superposition wasn't just an artifact of Heisenberg's strange-looking matrix mechanics. Schrödinger's theory, like Heisenberg's, had the major problems that haunt quantum physicists—including the concept of superposition. And the seemingly unavoidable idea of superposition so bothered Schrödinger that he came up with a thought experiment to show just how stupid the whole concept was. In the process, he threatened to tear down the whole edifice he and Heisenberg had built.

Schrödinger's thought experiment started with a quantum object in superposition—it doesn't matter what kind. It could be any binary choice: it could be a spin-up/spin-down atom or a photon that is polarized vertically and horizontally at the same time, anything that forces an object to choose between two alternatives, **0** and **1**. But for this example, let's say that it's an electron that hits a beam splitter and takes two paths at the same time. Both paths lead to a box—a box with a cuddly little kitten inside.[10] Path A is a dead end; if the electron travels down that path, nothing happens: **0**. But path B leads to an electron detector. When an electron strikes the detector, the detector sends a signal to an electric motor, which trips a hammer. The hammer breaks a phial of poison inside the kitten's box, and the poor thing dies instantly: **1**. An electron down path A, a **0**, means that the kitten lives, while an electron down path B, a **1**, means that the kitten dies.

10. Luckily for physicists, PETA doesn't get *too* angry about thought experiments.

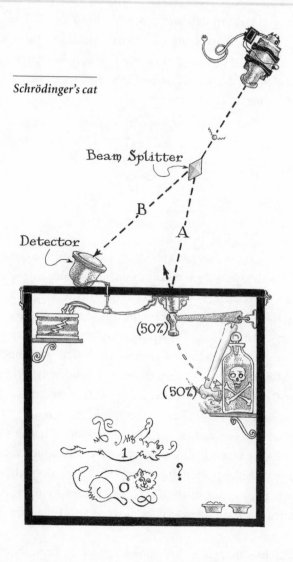

Schrödinger's cat

Beam Splitter

B

Detector

A

(50%)

(50%)

1

?

So what happens to Schrödinger's cat? Thanks to superposition, an electron will travel down path A and path B at the same time; it will be both **0** and **1** simultaneously. Therefore the electron both strikes the detector and doesn't strike the detector; the hammer both trips and it doesn't trip; the phial of poison breaks and it doesn't break, all at the same time. The kitten dies and it doesn't die: **0** *and* **1**. The laws of quantum mechanics imply that the kitten itself is in a superposition state; it is both alive and dead at the same time, in some weird ghost-like state of both life and death. How can something be both dead and alive at the same time?

(O & 1)

An alive-and-dead cat

But wait—it gets weirder. This state of superposition can exist so long as nobody opens the box. The moment someone extracts information about whether the cat lived or died, whether the cat is in state **0** or state **1**, it becomes equivalent to the case of a trip wire in an interferometer. When someone extracts information from the system, the superposition—in Schrödinger's terminology, the cat's wave function—collapses; the cat "chooses" life or death, and it suddenly picks one or the other, **0** or **1**. But, in principle, so long as the box remains undisturbed, as long as nothing has extracted information about the cat from the cat-and-box system, the cat's superposition remains undisturbed; the kitten is both alive and dead at the same time. It seems as if it's the act of observation, the act of extracting information, that kills the cat. Information can be deadly. This absurd-sounding conclusion seemed to be an unavoidable consequence of the principle of superposition.

When Schrödinger proposed his experiment, he knew the conclusion was silly. Macroscopic objects like cats don't behave like microscopic ones such as electrons, and it is laughable to think that something can be both living and dead at the same time. But if the mathematics of quantum theory says that this can happen, why don't we see half-living, half-dead people walking about on the streets?

(Graduate students don't count.) What prevents us from seeing superposition in large objects like baseballs and cars and kittens and people?

Some physicists have proposed that there's something fundamentally different that divides the quantum world from the classical one; some have speculated that there's some special size limit where, for one reason or another, the laws of quantum mechanics stop working and the classical ones take over. As far as experimentalists can tell, though, there is no such barrier. Scientists have been getting ever-bigger objects into a state of superposition. For example, the physicist Anton Zeilinger at the University of Vienna has been making Schrödinger-cat equivalents out of large molecules known as fullerenes. These fullerenes are round cagelike molecules that consist of sixty or more carbon atoms. By quantum standards, these things are absolutely gargantuan. Nevertheless, when Zeilinger shoots fullerenes through a grating, each one takes multiple paths on the way to a detector. Even though these objects are quite a bit larger than atoms and electrons and photons, they can be forced into taking two paths at the same time: they are forced into superposition. So far, scientists haven't found any size limit to the laws of quantum theory; everything in the universe should be subject to those laws.

So we have come to a paradox. The mathematics of quantum theory seems to imply that scientists should be able to put even large objects like baseballs and cats into superposition. But it is absurd to think that baseballs can be in two places at once and a cat can be both living and dead at the same time. If the laws of quantum mechanics apply to macroscopic objects, why don't macroscopic objects behave like quantum ones? It makes little sense.

Such a violation of common sense should sink a theory, but it was not the only such paradox. Einstein himself found one. Einstein hated the ugly, seemingly contradictory qualities of quantum mechanics and tried, over and over, to destroy the theory he helped create. He almost succeeded.

————

The paradox of Schrödinger's cat became one of the classic and trou-bling puzzles of quantum theory. However, as strange as superposition was, it was not the most troubling element of quantum mechanics—at least to Albert Einstein. Einstein saw a threat from another direc-tion; it seemed to threaten his hallowed rule that no information travels faster than the speed of light and looked as though it would allow scientists to build time machines.

Einstein and two colleagues, Boris Podolsky and Nathan Rosen, found this problem in 1935. They, like Schrödinger, created a thought experiment that would expose the absurdity of quantum mechanics, and it was a doozy. In it, they exploited a quantum-mechanical prop-erty now known as *entanglement*, which, coupled with the principle of superposition, threatened to reduce the entire framework of quantum mechanics to a bundle of contradictions.

The Einstein-Podolsky-Rosen thought experiment begins with a particle floating gently through space. All of a sudden, the particle decays—as particles are wont to do—into two smaller particles that fly off in opposite directions. If the two particles have the same mass, then they must zoom off with equal and opposite speeds; Newton's laws say as much. If one particle is heavier than the other, the heavy particle must be moving slowly and the light particle must be mov-ing fast.[11]

For argument's sake, let's say that our original particle decays into a heavy particle and a light particle, which shoot off to the left and to the right. Until you measure one or the other of the pair of particles created by the decay, you are left with a binary question. Either the light, fast particle has zoomed off to the left or it has zoomed off to the right. Or, looking at it another way, the leftward-moving particle is either light or heavy, fast or slow, 0 or 1. This particular set of particles is known as an EPR pair, after Einstein, Podolsky, and Rosen.

11. It's a consequence of the law of conservation of momentum.

Now, say that you measure the speed of the leftward-moving particle of the EPR pair: it is either fast or slow, light or heavy, **0** or **1**. As soon as you measure the speed of the left particle, you know which of those two states the left particle is in; you know whether it is moving fast or slow. But by measuring the left particle, you also get information about the *right* particle. If you measure the left particle and find it moving fast—it's a **0**—you automatically know that particle B is moving slowly—it's a **1**—and vice versa. A single measurement yielding a single bit of information—a **0** or a **1**—tells you about the state of *both* particles. The two are information-theoretically linked. This is entanglement. A single measurement of one of a pair of entangled particles gives you information about the other, without your needing to make any measurement whatsoever on the second particle. In an information-theoretic sense (and in a quantum-mechanical sense) the two objects behave, in a manner, as if they were a single particle. Measure one and you are really measuring both.

You can make an EPR pair whose particles are entangled in other ways. For example, you can create a set of particles whose spins are equal and opposite, just as you can create two whose speeds are equal and opposite. Measure one particle in such an entangled pair and find out that it is spin up and you will instantly know that the other is spin down. You can make a pair of photons, a pair of light particles, whose polarization is equal and opposite; if you know that the left-moving photon is polarized in the horizontal plane, the right-moving photon is polarized in the vertical plane.

So far, this isn't terribly weird stuff. This sort of thing happens in the macroscopic world all the time. For example, I can tell you that I put a penny in one box and a nickel in the other; when you open one of the boxes and see, say, a nickel, you know that the other box must contain a penny. One measurement, yielding one bit of information, tells you what "state"—penny or nickel—both boxes are in. However, unlike with pennies and nickels, you can throw superposition into the mix. When you entangle quantum particles that are in superposition, things get very, very hairy.[12]

As before, let's create an EPR pair of particles. For simplicity's sake, let's use spin instead of mass. We create the pair so that the two particles have equal and opposite spins: if one is spin down, the other is spin up; if particle A is in state **0**, particle B is in state **1**. But since we are dealing with quantum objects, we don't have to force each particle to commit immediately to state **0** or state **1**. We can set up the particles so that they are in superposition of **0** and **1**, spin down and spin up at the same time. And this really screws things up.

Like Schrödinger's cat, neither particle has "chosen" whether to be spin up or spin down. Each is both at the same time, so long as the particles remain undisturbed. They can speed away from each other for years and years, eventually winding up in different galaxies, each in that undisturbed state of superposition.

But what happens when you measure the spin of one of the particles? All of a sudden, the particle that was both **0** and **1**, spin up or spin down, "chooses" one of the states. When you extract information from one of the particles, the indeterminate superposition instantly collapses, and the state becomes, say, spin up, a **1**. As before, the act of measurement changes the state of the particle, changing it from a mix of **0** and **1** to a pure state of **1**. But this is much more troubling than a mere Schrödinger's cat, thanks to entanglement. Since the two particles are entangled, when we measure particle A and determine that it's a **1** we instantly know that the other particle must be a **0**. Since one bit of information tells us about *both* particles, extracting information from particle A is tantamount to extracting information from—measuring—particle B as well, even if it is halfway across the universe. The instant the act of measuring particle A causes it to change from a

12. Actually, even without superposition, entanglement *does* cause similar problems. The original EPR paper showed a potential problem because knowing a particle's momentum and position simultaneously would contradict Heisenberg's uncertainty principle. The superposition-plus-entanglement formulation is a later refinement of Einstein's argument devised by the physicist David Bohm.

mix of **0** and **1** states to a pure **1** state, the same measurement causes particle B to change from a mix of **1** and **0** states to a pure **0** state. As soon as we make one of the particles "choose" to be spin up, the other must, at the same instant, "choose" to be spin down. Somehow, as particle A "chooses" to be spin up, its twin, billions of light-years away, must instantly make the equal and opposite choice—its superposition must collapse. And you cannot explain this effect in classical terms; you cannot cop out by saying that the particles secretly "chose" their fate long before the measurement occurred. If you want, you could do a Monroe-Wineland–type experiment on the particles proving that they are in a superposition of two states rather than in a pure **0** or **1** state up until you make a measurement. It is a consequence of the mathematics of quantum theory: particle B "chooses" to be a **0** the very instant that you measure particle A and it "chooses" to be a **1**, not one moment before.

How can particle B make that choice instantly, even if it's far, far away in a distant galaxy? At first glance, it seems that there's no way it could. Since it would take a light beam billions of years to travel from our particle to its twin, and information can only travel as fast as the speed of light, it would seem that it would take billions of years for the distant particle to become aware of the measurement and particle A's choice, and only then could it collapse and "choose" the opposite state. But this is not the case. The particle knows instantly that its twin has been measured and knows what choice it makes. There is no time delay before particle B is "aware" that particle A has chosen **1** and itself collapses into state **0**. Einstein was horrified by this instant communication, this "spooky action at a distance." And yet, it has been verified.

In 1982, the physicist Alain Aspect saw this spooky action at a distance for the first time, and the experiment has been replicated many times since. Nowadays, the most advanced realizations of the EPR experiment take place at the University of Geneva, where Nicolas Gisin and his colleagues have been violating common sense with entangled particles for years. The particles they entangle are photons;

they create entangled pairs by zapping a crystal made of potassium, niobium, and oxygen with a laser. When the crystal absorbs a photon from the laser, it spits out two entangled particles that zoom away in opposite directions, which then are piped into glass cables.

Gisin's team has access to a large fiber-optic network that runs around Lake Geneva and into nearby towns. In 2000, the team shot entangled photons to the nearby villages of Bernex and Bellevue, which were more than six miles away from each other. By taking measurements with, effectively, an incredibly precise clock, they were able to show that the particles behaved in the manner Einstein predicted: the two were in superposition and always seemed to conspire to have equal and opposite properties upon measurement. And because of the distance between the two towns, there wasn't sufficient time for a light-speed "message" from particle A ("Help! I've been measured and have chosen to be a **1**. You know what to do.") to reach particle B before it, too, was measured and shown to be in the opposite state. In fact, the scientists determined that if some sort of "message" was sent from particle A to particle B, it had to travel at ten million times the speed of light to be received in time for particle B to "choose" its state before it, too, was measured. So, in a sense, the speed of quantum entanglement is (at least) millions of times the speed of light.

If these particles "communicate" faster than light speed, can they be used to transmit a faster-than-light message? If we had some source of entangled particles halfway between Alice and Bob, shooting a stream of particles in opposite directions, could Alice manipulate her part of the stream—encoding a bit on the particles—and could Bob at the other end receive that bit?

This question has been answered, as the next chapter will reveal. Nevertheless, the mystery about entanglement remains. It is as spooky and confusing as the day Einstein proposed it. Indeed, the two great mysteries of quantum mechanics are superposition and this spooky action at a distance. Why can microscopic objects be in two different places at the same time, and why do they have different properties from those of macroscopic objects? How can particles communicate with each other—instantly, even if they are halfway across the universe—and can they be used to transmit a message? These paradoxes go to the very heart of quantum theory; solve them and you have unraveled the mysteries of the quantum world.

Scientists are almost there. They have a theory that explains both of these paradoxes. This new idea is built upon the foundations that support both relativity and quantum theory: it is a theory of information more advanced even than Shannon's theory. It is a theory of quantum information.

CHAPTER 7

QUANTUM INFORMATION

What kind of liberation would that be to forsake an absurdity which is logical and coherent and to embrace one which is illogical and incoherent?

—James Joyce, *Portrait of the Artist as a Young Man*

The name Waterloo conjures up images of a great battle. In 1815, near Waterloo, Belgium, the Duke of Wellington defeated the forces of Napoleon Bonaparte. Nearly two centuries later a different Waterloo altogether—Waterloo, Canada—is the site of another battle. It is a battle for understanding. Ray Laflamme and his colleagues are trying to defeat the mysteries of the quantum world.

In his university lab, about an hour and a half from Toronto, Laflamme has two white, person-height, three-legged cylinders. They are not pretty to look at. They seem as though they would be more at home in an oil refinery or an industrial site than in a cutting-edge quantum lab. Yet these cylinders are more powerful tools for understanding the subatomic world than any conventional microscope could ever be.

Before you approach one of the cylinders, you must remove your wallet. Get too close to them and they will instantly lobotomize your credit cards, for the cylinders are tremendously powerful magnets. At a distance of more than a meter, they make paper clips or Canadian coins cling to one another, held together by the invisible influence of the magnetic field.

These magnets force atoms to dance. The powerful fields align atoms, making their spins line up and compelling them to twirl and twist in a complicated ballet of logic. Stored in those atomic spins is information—quantum information—and the complicated dance is a rudimentary computer program. These magnets and the atoms they affect make up a primitive *quantum computer*.

Just as computers manipulate information, quantum computers manipulate quantum information—an extension of Shannon's idea that takes into account the subtle mysteries of the laws of quantum theory. Quantum information is much more powerful than ordinary information; a quantum bit has additional properties that are unavailable to the classical 1s and 0s of Shannon information: it can be split into many parts, be teleported across a room, do mutually contradictory operations at once, and perform other seemingly miraculous feats. Quantum information taps a resource in nature that mere classical information cannot reach; because of these additional properties, a sufficiently large quantum computer would be able to crack all the cryptographic codes used for security on the Internet, and it would be able to perform feats of calculation utterly impossible for ordinary computers.

But more important, quantum information is the key to unraveling the mysteries of the quantum world, and quantum computers are giving scientists access to a previously unexplored realm. Because of them, experimentalists and theoreticians are beginning to divine the secrets that quantum information holds. And they are realizing that quantum information is more closely tied to the fundamental laws of physics than classical information is. Indeed, quantum information may be the key to understanding the rules of the subatomic world and the macroscopic world—the rules that govern the behavior of quarks and of stars and galaxies and of the universe itself.

———

Just as the era of classical physics gave way when the theories of relativity and quantum mechanics were born, so, too, the era of classical information theory has yielded to a broader, deeper theory known as quantum information theory. The study of quantum information is just beginning. Because the rules that govern the behavior of atoms and electrons and photons are very different from the standard Newtonian, classical, rules of lamps and balls and flags and other macroscopic objects, the information that is carried by a quantum object like an electron is different from the simple bits that can be inscribed on a classical object. While classical information theorists talk about information in terms of bits, quantum information theorists talk about information in terms of quantum bits, or *qubits*.[1]

In classical information theory, the answer to any yes/no question can always be answered, well, as a yes or a no—a **1** or a **0**. But with quantum theory, this nice, easy distinction between yes and no broke down. Quantum objects can be two things at once, on the left side of an interferometer or the right, spin up and spin down, both **1** and **0** at the same time. While a classical object can never be in an ambiguous superposition of two states—it must always be in one state or another, on or off, left or right, a **1** or a **0**, but not both at the same time—a quantum object can be.[2] So, even with a nice clear-cut yes/no question (Is Schrödinger's cat alive?), there is often no way to answer it with a straight **1** or **0**: a cat can (theoretically) be both alive and dead, an electron can be on both the left and right, and light can be both particle and wave. Mere **1**s and **0**s cannot capture duality or superposition; the realm of the quantum object doesn't have the neat dichotomies of the classical world.

But, as we saw, quantum theory (and relativity, for that matter) is a

1. There are also more complicated quantum creatures known as qutrits and qunits, but in this book qubits will suffice.
2. Remember, superposition is *not* equivalent to being somewhere in between the two states; a dim lightbulb or a billiard ball in the middle of a table rather than on the left or the right is still in an unambiguous position describable in terms of classical bits. A quantum object is not in a definite state like this; it takes on two values simultaneously, and so is in a state of superposition, taking contradictory values at the same time.

theory that deals with the transfer of information. So how can scientists talk about information on a quantum object if the 1s and 0s of classical information theory are insufficient to describe what's going on? That is where the qubit comes in. Quantum bits, unlike classical bits, can take on two (or more) contradictory values at the same time. They can be both 0 and 1 simultaneously. While you can't describe the live *and* dead state of Schrödinger's kitten with a classical bit, you can do so with a qubit. But to get into the nature of quantum information, I have to introduce a new notation for qubits that captures the quantum nature of quantum information.

A classical cat can only be alive, 0, or dead, 1. But an ideal Schrödinger's cat can be both alive and dead at the same time: (0 & 1). That is a qubit, both 0 and 1 at the same time. The cat in superposition can theoretically maintain this live *and* dead state—it can store this (0 & 1) qubit—so long as nobody looks in the box, but as someone tries to determine whether the cat died or survived, the superposition collapses. The (0 & 1) state instantly changes to a classical bit. The cat "chooses" either a 0 state—the cat lives—or a 1 state—the cat dies.

The notation for qubits is a little cumbersome, but it's necessary.[3] A qubit is not the same as one or two classical bits; (0 & 1) is very different from a 0 and a 1, as we shall soon see.

Just as it doesn't matter what medium a classical bit resides in—you could store a bit in a flashlight bulb or a flag or a punch card or a piece of magnetic tape—it doesn't matter what object a qubit is written on. The qubit can represent the position of an electron in an interferometer: left is (0), right is (1), and the superposition of left *and* right is (0 & 1). It can represent the orientation of an atom's spin: up, down, or up *and* down. It can represent the polarization of a photon: vertical,

---

3. Actually, scientists use what is known as bra-ket notation to describe an object's quantum state. A *bra* is a mathematical object that is symbolized like so: < |; a *ket* is a closely related mathematical object that is symbolized like so: | >. A Schrödinger's cat in a state of superposition can be written in kets as |0> + |1> (divided by the square root of 2, for technical reasons). Why use kets? Long story . . . but have you ever tried to get a cat into a bra?

horizontal or vertical *and* horizontal. What is important is not the medium that the qubit is on, but the quantum information the qubit represents.

The paradox of Schrödinger's cat hinges upon the difference between qubits and regular bits. An atom or an electron or any other quantum object can be put into a state of superposition, taking both a left and right path at the same time. In essence, this stores a qubit on the atom: it is in a **(0 & 1)** state, rather than a pure **(0)** or a pure **(1)** state. When the atom enters the box, it transfers this qubit to the cat. The cat enters a **(0 & 1)** state just as the atom was in; the only difference is that **(0 & 1)** no longer represents a superposition of "left path" and "right path" as it did with the atom. With the cat, **(0 & 1)** represents a superposition of "living" and "dead." The *form* of the information changed, but the information itself, the qubit **(0 & 1)**, remained the same.

Quantum information theory—the study of qubits—is a hot area in physics right now. On the practical side, qubits can do things that classical bits can't. Machines that manipulate qubits, quantum computers, can do things that are impossible for a classical computer. Quantum computers are, theoretically, more powerful than classical computers could ever possibly be. Build a sufficiently large one and you can crack all the ciphers on the Internet; as a little game, you could listen in to any "secure" online transaction, crack the code, and steal the credit-card numbers and personal information that have been exchanged—something that is beyond the ken of the best supercomputers in the world. It is no coincidence that the U.S. Department of Defense is paying close attention to developments in quantum computing, and the potential power of quantum computing is why quantum information theorists are having little trouble finding money to do research. Yet practical applications are not why many scientists are interested in quantum computers: they see quantum computation as a way of understanding the paradoxes of quantum mechanics, as will become apparent shortly.

Whatever the reasons people give for studying quantum computation, before you can start cracking codes or creating quantum paradoxes in your lab, you have to be able to manipulate and store qubits. This means you need to procure quantum objects on which to store your quantum information. In Ray Laflamme's lab, physicists use atoms in a fluid. A molecule such as chloroform has a number of carbon atoms in a row—a number of carbon nuclei surrounded by clouds of electrons. Each carbon nucleus has a spin associated with it. Ordinarily, these spins point any which way, but with a powerful magnet they try to line up along with the magnetic field. Laflamme uses this tendency to make the atoms point any way that he desires. He can force a nucleus to point up or down, storing a **(1)** or a **(0)**, or he can put it into a superposition, a **(0 & 1)** state, by combining a carefully timed series of pulses of radio waves with the magnetic field. If you measure this superposed state, trying to determine what state it is in, 50 percent of the time it will collapse and give you a **0** reading, and 50 percent of the time it will collapse and give you a **1** reading, but until you take that measurement, it is in a superposition, just like Schrödinger's cat.

Not all superpositions are quite that simple. Just as there are biased coins—say, one that comes up tails 75 percent of the time—there are biased superpositions. Imagine, in our original Schrödinger experiment, that we add an additional layer to our interferometer. As before, we send our electron through a beam splitter, which puts the electron in a superposition of two possible states: left and right. It is in a **(0 & 1)** state. But this time, each of those paths leads to *another* beam splitter. The electron now is in a superposition of four states. It is simultaneously in four different possible places: path A, path B, path C, and path D, and if we make a measurement, there is a 25 percent chance that we measure the electron on any given path. If only path D leads to the trigger that breaks the poison phial, then there is only a 25 percent chance that the cat dies; 75 percent of the time it will survive—when we open the box. But until we open the box, the cat is in a state of superposition of 75 percent life and 25 percent death; it is in a quantum state that we can represent by ([75%]**0** & [25%]**1**). That is, the cat is simultaneously alive and dead; it is in a superposed state, but like the biased coin, when we measure the state the cat is in it is three times more likely to wind up alive than dead. When you measure the cat, three times out of four you will find out that it is a **0**, alive; one time out of four it is a **1**, dead.

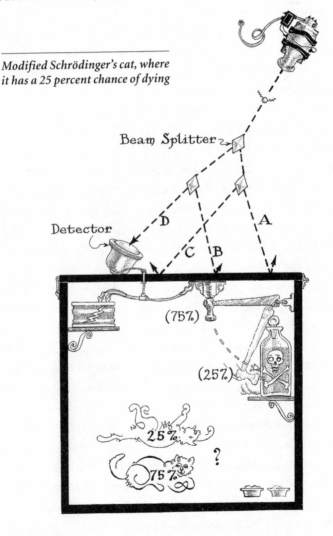

Modified Schrödinger's cat, where it has a 25 percent chance of dying

This is a superposition just as valid as the one in the original Schrödinger's cat experiment; it just has slightly different probabilities for the outcomes of a measurement. In fact, there are lots and lots of possibilities for different sorts of superpositions with different probability outcomes. If you set up your experiment right, you can manipulate a superposition so that the cat lives *x* percent of the time and dies *y* percent of the time: it is in an ([*x*%]**0** & [*y*%]**1**) state, where *x* and *y* add up to 100 percent. Note that what we called the simple (**0** & **1**) state before is really the ([50%]**0** & [50%]**1**) state. Since this particular unbiased state will appear and reappear so often in the book, I will still refer to it as the (**0** & **1**) state for brevity if the context allows.

Laflamme's enormous magnets can put atomic nuclei into whatever state the scientists wish. They can store any ([*x*%]**0** & [*y*%]**1**) qubit, for any *x* and *y* that they desire, upon the spins of those nuclei. Then they can manipulate those qubits with magnetic fields and radio waves just as a computer manipulates **0**s and **1**s with electricity. For example, a computer can negate a bit: if a bit starts off as a **0**, negation turns it into a **1** and vice versa. Laflamme, and dozens of other researchers, can negate a qubit. If you start off with a quantum object that stores the qubit ([*x*%]**0** & [*y*%]**1**), after negation that object will store the qubit ([*y*%]**0** & [*x*%]**1**)—a quantum negation. The researchers can do many other things to these qubits as well. Through various techniques (which depend on whether the qubits are stored in atomic spins, light polarizations, or other quantum properties), they can manipulate a qubit of quantum information in ways just as complex as the way a computer manipulates classical bits of classical information. In essence, researchers around the country have been building rudimentary quantum computers.

In 1995, the physicist Peter Shor proved that a quantum computer of this sort could factor numbers dramatically faster than any classical computer could, and this, in turn, would make most of the cryptography in use today instantly obsolete.

Public-key cryptography, which is the basis for most of the cryptography on the Internet, is a one-way street for information. It is like dumping a letter in a mailbox. Anyone can encrypt a message just as anyone can put a letter in a mailbox, but only the person with the right cryptographic "key" can decipher the message, just as only the postman can open the mailbox and retrieve all the information inside. Information checks in, but it doesn't check out—unless you know the secret key. Mailboxes are one-way devices because they rely upon a shape that makes it easy to let a letter fall into the box but very hard to stick your arm inside and pull a letter out. Public-key cryptographic systems are one-way devices because they rely upon mathematical functions that are easy to do and hard to undo. Such as multiplication.

For computers, multiplying two numbers together is very, very easy. They can multiply even huge numbers together in a matter of milliseconds. But the opposite—factoring—is very, very hard. If you choose a sufficiently large target number wisely, even the best classical computer in the world will never, ever, be able to figure out which two numbers, multiplied together, yield the target number as their product. This is the one-way device at the heart of most public-key cryptography; the difficulty in factoring numbers is what makes these ciphers secure.

When Shor proved that a quantum computer could factor numbers many, many, many times faster than a classical computer could, his discovery went straight to the core of what makes public-key cryptography secure. A number that might take a classical computer the entire lifetime of the universe to factor might take a quantum computer only a few minutes. Shor's algorithm drives the wrong way down that one-way street, making it easy to factor numbers as well as multiply them together. When factoring numbers becomes easy to do, public-key cryptography becomes useless. And at the heart of Shor's algorithm is the weirdness of quantum information: qubits make possible things that are impossible for classical computers.

Factoring extremely large numbers is not the only "impossible" thing a quantum computer can do. Quantum computers can run roughshod over many of the hallowed dicta of classical information theory. Remember the "guess a number" game in chapter 3? If I am thinking of a number from 1 to 1000, you can always guess it by using a series of ten yes/no questions. Classical information theory says that you need ten of those yes/no questions to be 100 percent certain of guessing the number correctly; you need ten bits of information to completely wipe out the uncertainty about which number I am thinking of. In 1997, the physicist Lov Grover at Bell Labs in New Jersey proved that a quantum computer with ten qubits of memory can do the same task with *four* yes/no questions. The difference between the quantum and the classical gets more profound as the problem gets bigger: a classical problem requiring 256 yes/no questions gets solved in a mere sixteen yes/no queries in a sufficiently large quantum computer.

Shannon would have thought Grover's algorithm to be impossible. Since information theory pares down a question to its incompressible essence, it should be impossible to answer a 256-bits-of-information question with a mere sixteen yes/no queries. But Grover's algorithm does just that. To see how, let's take a somewhat smaller problem. We've got a combination lock with sixteen possible combinations, 0 through 15. Only one of them (say, 9) is the correct combination that will open the lock.

In classical information theory, we would need to ask four yes/no questions about the combination to figure out which combination will work. Here are four questions that would suffice. Question 1: Is the correct combination odd? Nine is certainly odd, so the answer is yes: **1**. Question 2: Divide the combination number by 2 and round down to a whole number. Is *this* number odd? Nine divided by 2 is 4.5; rounded down, it is 4. So, no, the answer is **0**. Question 3: Do the same thing again: divide by 2 and round down. Is *this* number odd? Four divided by 2 is 2, an even number. So the answer is no: **0**. Question 4: Once more, do the same thing. Is *this* number odd? Two divided by 2 is 1; 1 is odd, so the answer is yes: **1**. Four questions, four answers, and there is only one possible number in the range of 0 through 15 that satisfies all four answers: 9. (Math-savvy readers might have noticed that our questions reduced the number 9 to binary code: **1001**.) Only after these four questions—or four questions of a similar ilk, such as ones of the higher/lower variety—would we know that the correct combination is 9 and be able to use it to open the lock.

Grover's algorithm, however, takes an entirely different approach. Essentially, using the principles of superposition and entanglement, it asks all the questions at once, rather than one at a time. More specifically, Grover's algorithm uses four qubits, each of which starts out in a balanced superposition: ([50%]**0** & [50%]**1**). But the four are linked via entanglement. It is as if the four qubits form one large object. This is kind of messy looking, but what we've got is an object in the superposed state:

[([50%]**0** & [50%]**1**) ([50%]**0** & [50%]**1**) ([50%]**0** & [50%]**1**) ([50%]**0** & [50%]**1**)]

If we were to take a measurement right now, the first qubit has an even-odds chance of being a **0** or a **1**; likewise for the second, the third, and the fourth qubit. In essence, we have got sixteen different possible outcomes, all superposed on top of one another: **0000, 0001, 0010, 0011, 0100, 0101, 0110, 0111, 1000, 1001, 1010, 1011, 1100, 1101, 1110, 1111**. In binary code, this is just all the numbers from 0 through 15, all superposed.

The next step in Grover's algorithm is the mathematical equivalent of forcing that horrible superposed object into the combination lock. Essentially, it asks a yes/no question: Does this four-qubit thing fit? The answer is received in a form that doesn't immediately reveal the combination of the lock, but the act of cramming does have an effect on the qubits; the probabilities get changed so that the superposition is no longer 50:50. The incorrect answers get less probable, and the correct ones get more probable. In our case, where the correct combination is 9, or 1001 in binary, our four qubits coming out of the lock might look like this:

[([25%]**0** & [75%]**1**) ([75%]**0** & [25%]**1**) ([75%]**0** & [25%]**1**) ([25%]**0** & [75%]**1**)]

Pass this mess through the lock one more time; the correct answers are enhanced and the incorrect ones are diminished, yielding:

[([0%]**0** & [100%]**1**) ([100%]**0** & [0%]**1**) ([100%]**0** & [0%]**1**) ([0%]**0** & [100%]**1**)]

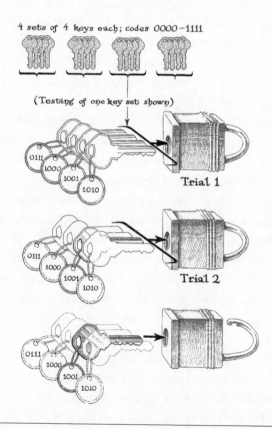

The Grover algorithm: the incorrect combinations quickly disappear

For a classical problem requiring n questions, you need to do $\sqrt{n}$ passes to get to this point. The incorrect answers are wiped out, and the correct ones are all that is left. If you make a measurement of the four qubits, the superpositions will collapse to give you **1001**—the combination to the lock. Grover's algorithm only asked a yes/no question twice: Does this four-qubit thing fit? But because that "four-qubit thing" was in superposition, it was actually asking the yes/no question about many combinations simultaneously. It takes a little mathematical massaging to ensure that the right answer comes out—hence the two passes. Nevertheless, the quantum computer was able to ask fewer yes/no questions than would be classically needed.

For a problem that requires four bits of information, the Grover algorithm gets the answer in two questions—an improvement, but nothing spectacular. But for big problems, such as those requiring 256 bits of information or more, the difference in the time it takes to ask $\sqrt{n}$ questions compared with asking n questions is enormous. It could mean the difference between a few seconds of computation and needing the most powerful computer to work from the beginning of the universe to its end before getting the correct answer.

Shor's factorization algorithm uses qubits in a similar way. It effectively tests many, many numbers in superposition, all at the same time. A set of qubits in superposition, all entangled together, allow you to test zillions of combinations at once. It's as if you have the master key to all the cryptographic locks in the universe. Quantum information is immensely powerful—but scientists are having trouble harnessing it.

In 1998, the first quantum computer was born. Isaac Chuang and Neil Gershenfeld, physicists at IBM and MIT respectively, used a setup just like Ray Laflamme's as the heart of their computer. The computer itself was made of atoms in a powerful magnetic field; the qubits were the spins on those atoms. By carefully manipulating the magnetic fields, Chuang and Gershenfeld made the atomic spins do a dance that corresponded to Grover's algorithm. The atoms twisted and bounced, and after one pass the two-qubit quantum computer correctly picked the target number from among four possible choices. It had done something that was classically impossible.

But quantum computing research moves very slowly. In 2000, Laflamme announced that he had created a seven-qubit quantum computer, and in 2001 Chuang used a similar seven-qubit computer and Shor's algorithm to factor . . . the number 15. Into 3 and 5. Something that most ten-year-olds can do without a moment's hesitation.[4] And yet it was a great milestone for quantum computing: it was the first time that anyone had been able to run Shor's algorithm.

The problem is that to break the codes on the Internet, you need a quantum computer that uses several hundred qubits, all linked together by entanglement. Scientists are struggling to get to ten qubits right now. It is generally agreed that the technique that Laflamme and Chuang and Gershenfeld and others use won't scale up very much further.[5] Engineers will have to turn to other techniques to make their quantum computers; they will have to store their qubits on other media besides atoms in a strong magnetic field. But each medium that they have tried—the polarization of light, the charge on a silicon trap called a quantum dot, the direction of current in a tiny loop of wire— has drawbacks that make it difficult to create a whole bunch of qubits that are entangled with each other. None of these techniques is currently as advanced as the atomic-spin quantum computer.

By comparison, even the earliest commercial classical computer, UNIVAC, had tens of thousands of bits of memory. As neat as it is, a

4. And in fact, the achievement was *less* impressive than it seemed. It took advantage of the fact that 15 is 1 less than 2^4, saving a little memory in the process.
5. In fact, this problem is related to a controversy about whether the atomic-spin computer is truly a quantum computer, and what makes a quantum computer "quantum," but that's a huge can of worms. The important thing is that these computers are doing quantum algorithms with quantum information.

quantum computer with seven or ten qubits is not much use at all—to codebreakers. It's not certain that scientists will ever be able to build a quantum computer large enough to crack commercial codes. Nevertheless, scientists are thrilled to be playing around with their tiny quantum computers, and it has to do with the real reason that scientists are so interested in quantum information theory. Codebreaking is fun and important, but it is nothing compared to the questions that scientists are asking of Nature. When physicists manipulate even a single qubit, they are trying to understand the nature of quantum information. And by understanding quantum information, they are understanding the substance of the universe—the very language of Nature.

The reason quantum information theorists are so excited about their field concerns the paradoxes of quantum mechanics. It turns out that these paradoxes are all, at their cores, paradoxes about information storage and transfer.

For example, the paradox of Schrödinger's cat comes from trying to store a qubit on a classical object. For some reason, you cannot store a qubit on a cat; something prevents big, classical, sloppy objects like cats from being used as media for qubits. Cats can store classical bits just fine, though keeping track of 1s and 0s by killing or letting survive a series of cats rapidly gets expensive. But when you try to store a qubit, a **(0 & 1)**, on a cat, you get Schrödinger's absurd paradox. Something weird happens when you try to transfer a qubit from a quantum object to a classical object—from, say, an electron to a cat.

Similarly, Heisenberg's uncertainty principle is a problem of information transfer. When you measure a particle's property—say, an atom's position—you are transferring information from a quantum object (the atom) to another one (such as the equipment that records your atom's position). Yet the mathematics of quantum theory says that you cannot gather information about two complementary attributes of a quantum object at the same time. You can't know a particle's position and momentum simultaneously, for example. The act of measurement, of transferring information from the particle to you, affects the system you are measuring. When you gather information about a particle's position, you lose information about its momentum.

The weirdness of entanglement is also a problem of information transfer. When you measure one particle in an EPR pair, you are getting information about both of the particles; it seems as if you're transferring information from a very distant object to your measuring device at speeds greater than light speed. And since the act of information transfer affects the particle you're transferring information from, it seems as if you are instantly manipulating a particle halfway across the universe. What is the nature of the connection between two entangled particles? How can two objects "conspire" to remain entangled even when there is no way they can exchange information even at light speed?

Though most scientists believe that the laws of quantum theory should apply to everything—to cats as well as to atoms—macroscopic objects clearly don't display quantum behavior the way microscopic ones do. If they did, if classical objects behaved like quantum ones, quantum theory would not seem so alien; we would be used to it. But quantum mechanics *is* alien—it's downright absurd—and the central element in all of that absurdity is the act of transferring quantum information. Whenever you perform a measurement and gather information about a quantum object, or whenever you transfer quantum information from an atom or a photon or an electron to another object, things are likely to get bizarre.

In fact, all the absurdity of quantum theory—all the seemingly impossible behaviors of atoms, electrons, and light—has to do with information: how it is stored, how it moves from one place to another, and how it dissipates. Once scientists understand the laws that govern these things, they will understand why the subatomic world behaves so differently from the way the macroscopic one does, why cats can't exist in a superposition of life and death while atoms can be in two places at once. They will understand why an entangled EPR pair of particles can "sense" each other's choice halfway across the universe even though people can't read each other's minds at a distance. And though most scientists believe the laws of quantum theory apply to big objects as well as small ones, there is clearly a difference in the way macroscopic and microscopic things behave. These are the fundamental questions of quantum theory, and they have obsessed scientists since the 1920s.

The answers to these questions may now be within reach, and this is why quantum information theorists are spending their time manipulating a mere handful of qubits. Though quantum computers are a long way from cracking codes and factoring numbers, they are still incredibly powerful. Scientists can use them to understand the way quantum information behaves; they can store quantum information, transfer it, measure it, and watch it dissipate. The real value of quantum computers is not in the programs they can run, but in the knowledge they are giving scientists about the way the quantum world works—and even a single qubit can reveal the rules that govern the transfer of quantum information. Indeed, the simple act of measuring a quantum object is the crux of the quantum dilemma, and that simple act has very strange effects.

One such effect seems a little arcane at first, but upon a little reflection it's quite troublesome. You can keep a radioactive atom from decaying simply by looking at it—by measuring it. This goes against commonsense wisdom about how radioactive atoms behave.

A radioactive atom has an unstable nucleus. For example, uranium-235 is quivering with energy, trying to break itself apart. However, the binding force that keeps the neutrons and protons tied together manages to keep that energy in check—for a while. At some random time, the nucleus breaks into two big pieces and releases a lot of energy. For decades, scientists measured the rate at which those nuclei broke apart, or *decayed*. If left to their own devices, uranium atoms would do it at exactly the same rate. Each radioactive nucleus has a characteristic rate at which it decays. This rate is expressed as the nucleus's *half-life*, and it is a fundamental property of each radioactive nucleus. Leave a bunch of uranium atoms alone in a jar and after a certain time a predictable number of them will have fallen apart. It seemed that nothing you could do would prevent those atoms from decaying.

Look at the nucleus in a slightly different way and a loophole appears. From the perspective of quantum theory, each unstable nucleus is actually a Schrödinger's cat; it is a nucleus that is constantly in a state of superposition. One quantum state, (0), is the nucleus as an unbroken if unstable whole: in state (0) the nucleus is a single, undecayed object. The other quantum state, (1), is the nucleus decayed into two pieces. Usually, the atom starts off in a superposition that is heavily biased toward the (0) state—it might even start off in a pure (0) state—or, in the more cumbersome notation, the ([100%]0 & [0%]1) state. But as time passes, the bias changes. The nucleus's superposition gets more and more pronounced. As time passes, it will evolve to a ([99.9%]0 & [0.1%]1) state and then, say, a ([98%]0 & [2%]1) state and then, sometime later, an ([85%]0 & [15%]1) state. At some point, when the probability of the (1) state is high enough, the superposition spontaneously collapses and the nucleus breaks apart. It is as if Nature measured the nucleus and the celestial coin flip decided that the nucleus was in the (1) state—the broken-apart nucleus—rather than the unbroken (0) state. (More on this spontaneous collapse shortly.)

But according to quantum theory, you can tinker with the decay of a nucleus simply by measuring it over and over and over again. If you start off with a pure ([100%]0 & [0%]1) state, the nucleus is unbroken. If you measure the nucleus as soon as the Superposition just begins to evolve, say, by bouncing a photon off it, you are nearly guaranteed to measure that it is in the (0) state. The nucleus has not had

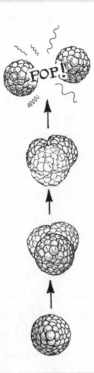

Nuclear decay from a quantum perspective

time for the superposition to evolve very far; it might be in the ([99.99%]**0** & [0.01%]**1**) state, so a measurement will almost always yield a **0**: the nucleus hasn't broken apart. But the act of measurement destroys the superposition. Measuring the nucleus drops it back to the ([100%]**0** & [0%]**1**) state once more; gathering information about the nucleus wipes out the superposition and puts the nucleus back in a pure state. You are back where you started. If you quickly measure the nucleus again, you reset the superposition once more. Quick, do it again. Again. Again. Each time, you are nearly guaranteed to see an unbroken nucleus, and each time, you reset the nucleus into its pure (**0**), unbroken state. Quick, repeated measurements prevent the superposition from ever evolving; the nucleus never really visits the **(1)** state at all, so there is almost no chance it ever decays. Keep measuring the nucleus over and over and you can prevent it from decaying. It's true: a watched pot never boils.

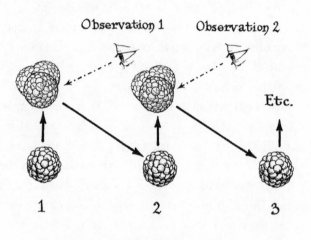

The quantum Zeno effect

Repeated measurements can prevent a nuclear decay. This effect, known as the *quantum Zeno effect*,[6] has been studied in labs using trapped ions and photons. And theorists suggest that just the opposite can happen: it might be possible to *induce* an atom to decay by watching it carefully. The quantum Zeno and quantum anti-Zeno effects show that the act of measurement—the transfer of information—is intimately related to a real, physical phenomenon like nuclear decay. Somehow, quantum information is tied to the laws that govern how matter behaves.

---

6. It is named after the philosopher Zeno of Elea, who argued that subdividing a footrace into infinite little segments makes it impossible to complete the race.

In fact, you can recast the physical process of nuclear decay entirely in the language of quantum information. Even absent a human observer, you can view the spontaneous splitting of an atomic nucleus as an act of information transfer. The nucleus starts out in a pure, unbroken state and evolves into a superposition of decayed and undecayed states; like Schrödinger's cat, it is both broken and unbroken at the same time. Then something happens. Something gathers information about the nucleus; something measures the state of the atom. Something transfers information about the state of the nucleus into the surrounding environment. This information transfer collapses the superposition; depending on the celestial coin flip, the nucleus "chooses" whether to be in a pure undecayed (0) state or a pure decayed (1) state. If the former, the process begins all over again; if the latter, then the nucleus spontaneously decays, just as radioactive atoms are expected to do from time to time. This picture of radioactive decay is completely consistent; you can use it to predict how many atoms will decay in a given time, and it will give you the right answer. You can view nuclear decay as an information transfer process, but one sticking point remains: the "something" that is doing the measuring. What is it that gathers information about the atom and disseminates it into the surrounding environment?

That something is Nature. Nature itself is constantly performing measurements. And this is the key to solving the paradox of Schrödinger's cat.

Scientists generally don't view Nature as a being of any sort. The vast majority do not believe that the universe is conscious. Nor do they believe that a supernatural creature is running around with a tiny calipers. But they absolutely believe that Nature—the universe itself—is, in a sense, continually making measurements on everything.

The universe is flooded with particles. The Earth is bombarded by photons from the sun, and it is thanks to those particles that you can perceive your environment so well. When you look out a window at a nearby tree, your brain is processing information that Nature has gathered for you. A photon from the sun has bounced off a leaf of the tree into your eye; the information about that tree would be there whether or not your retina was there to receive that information. The sunlight beating down on the tree is, in essence, a natural measurement: it takes information about the tree—the tree is sixty feet tall, and green, and swaying in the breeze—and sends this information into the environment.

Even if you close your eyes and ignore the information in those photons, you may still be able to perceive the tree. You can hear the wind rustling the leaves: you can sense the motion of air molecules, which bounce against the tree and against one another. These are what causes sound waves. The breeze takes information about the tree and sends it out into the surroundings. Whether your ear is there to perceive the rustling or not, that information has been disseminated into the environment.[7] Of course, you can measure the tree yourself. You can go up to it and touch it and feel the pressure of its bark molecules against your hand molecules, but you don't need to do that to know that the tree is there; you can process the information that Nature has already collected for you about the tree in the form of light and sound. The particles of light and particles of air are Nature's probes, Nature's measuring devices. You are simply receiving the information that has already been deposited on those particles.

Snuff out the sun and remove the Earth's atmosphere and those sources of information would no longer be available to you. (Though your lost sensations would hardly be your primary concern if the sun and atmosphere suddenly disappeared, of course!) You would no longer be able to perceive the tree through reflected light or through

---

7. Yes. It would make a sound if it fell. No question. Take that, Zen monks!

sound waves, as the Earth would be pitch black and airless. No human would be able to sense the tree from a distance, because the main portals for absorbing the information that Nature has gathered for us— our eyes and our ears—would no longer be receiving any signals. But that doesn't mean that Nature would have stopped its measurements just because humans weren't receiving any signals. Far from it.

Nature doesn't need the sun or the wind to make a measurement of the tree. Photons from distant stars are also bombarding the Earth, and though our eyes are too weak to perceive a tree only by starlight, a clever scientist with a photodetector could make out the outline of the tree—the information is still streaming into the environment. The Earth itself, since it is warmer than absolute zero, radiates photons as well. An infrared camera could pick up that radiation, and when it bounces off the tree it, too, reveals the tree's silhouette. (The tree, too, is radiating infrared radiation that contains information; we can only stop this by bringing the tree to absolute zero.) Even if a frozen tree were floating about in deep space, shielded from the warmth of the Earth and the wan light of distant stars, Nature still measures the tree. The universe is teeming with photons that were born shortly after the big bang; these, too, constantly knock and jostle the tree, gathering information about it and sending it into the environment. It's a simple trick to verify that the information is actually there: an observer with a properly tuned detector could spot the photons ricocheting off the tree.

Even without those photons, Nature still measures the tree. Space is saturated with cosmic rays from distant galaxies, as well as neutrinos— tiny, nearly weightless particles that seldom interact with matter— from the most distant reaches of the galaxy. These, too, pass through and bounce off the tree, and though it's technically very difficult, in theory a scientist armed with the proper detector could spot the way the tree affects the passing particles. The information is still disseminated into the environment.

What happens if we completely isolate the tree from the particles that suffuse the universe? What happens if we lock the tree in a vacuum in a box at absolute zero—so it doesn't radiate light—a box that shields it from neutrinos and cosmic rays and photons and electrons and neutrons and all the other probes that Nature uses to gather information? Would we be able to prevent Nature from getting information about the tree? Surprisingly, the answer is no. Nature always finds a way to gather information about the tree. Always—even in the deepest vacuum, and even at absolute zero.

Even if we shield the tree from all particles—from all the means that Nature uses to gather information—Nature creates its own particles at every point in space. On the smallest scales, particles are constantly winking in and out of existence. They appear, gather information, disseminate it into the environment, and disappear to the nothingness from whence they came. These evanescent particles are the vacuum fluctuations introduced in chapter 2, and they occur in every region of the universe, even in the deepest, coldest vacuum. Vacuum fluctuations make it impossible to shield an object completely from Nature's measurements. They were theorized (and then experimentally confirmed) as a consequence of Heisenberg's uncertainty principle.

As I explained in the last chapter, Heisenberg's uncertainty principle is a restriction upon information. No observer can simultaneously know, with perfect accuracy, two complementary attributes of an object at the same time. For example, it is impossible to have perfect information about a particle's position and its momentum at the same time; indeed, knowing everything about a particle's position means that you have no information about its momentum. But information is related to the state of physical systems. Information exists, whether or not somebody is extracting it or manipulating it; you don't need a human to measure a particle's quantum state for the particle to *have* a quantum state. Information is an inherent property of objects in the universe, and Heisenberg's uncertainty principle is a restriction upon information. Therefore, Heisenberg's uncertainty principle is actually a law about the quantum state of objects in the universe, not just about the measurement of that quantum state.

When most popular science books introduce Heisenberg's uncertainty principle, they talk about how a measurement "disturbs" the system being measured. Bounce a photon off an electron to measure its position and you give it a little kick of energy: you change its speed, reducing your information about the particle's momentum. But this is an incomplete description, because the uncertainty principle holds whether or not a scientist is measuring anything. It holds for all aspects of Nature, regardless of whether anyone's gathering any information. It holds even in the deepest vacuum.

Another one of the pairs of complementary attributes in quantum mechanics is energy and time. Know precisely how much energy a particle has and you don't have any clue about how long the particle has that energy, and vice versa. The rules of quantum theory say that this principle applies not only to particles but to *everything* in the universe—even a region of empty space.

Empty space? Doesn't empty space have *zero* energy? Well, no, according to Heisenberg's uncertainty principle. If it had exactly zero energy, then we would have perfect information about the energy in a region of space. By the energy-time complementarity, we would have *no* information about how long the region of space maintained that energy; it would have no energy only for an immeasurably short instant. After that, it *must* have some energy. Likewise, via the momentum-position complementarity, if we have a very accurate fix on a region of space—if we are looking at a very small region with very little uncertainty—we have little knowledge about how much momentum is in that region. As we zoom in to a smaller and smaller region (and thus, we are observing a region with greater and greater positional accuracy), we know less and less about the momentum in the region we are observing. Since a quantity of exactly zero momentum would mean that we have impossibly perfect information about the momentum in the region, there must be nonzero momentum in that region. Even in a vacuum.

That's downright weird. How can an empty region of space contain energy and momentum if there is nothing to carry that energy or momentum? Nature takes care of that for us: particles are constantly winking in and out of existence. They are born, carry energy and momentum for a brief moment, and then die. The more energetic the particles, the shorter they live, in general (thanks to the energy-time relationship), and the more momentum they carry, the smaller the region in which they live and die (thanks to the momentum-position relationship). In other words, even in the deepest vacuum, particles are created and destroyed, and the more you zoom in, the more of these particles there are, the shorter they live, and the more energetic they are. These particles constantly bump into things, gather information about the objects they encounter, spread that information into the environment, and disappear once more into the vacuum. These are vacuum fluctuations.

This is not a fanciful idea. It has actually been measured in the lab. Under the right conditions, these evanescent particles can move plates around, a phenomenon known as the Casimir effect. In 1996, physicists at the University of Washington measured the force exerted by these vacuum fluctuations. Though the force is very tiny—about 1/30,000 of the weight of an ant—they managed to confirm that the particles were, indeed, exerting this force. A number of other experiments since then have confirmed the Washington result. These evanescent particles exist; we can even see the effects they have. And since particles are constantly winking in and out of existence in every region of space, Nature is always making measurements with these particles. It's impossible to prevent it from doing so.

Thanks to these fluctuations, sudden, spontaneous collapses of a superposition—such as what happens in a nuclear decay—make sense. You need not have human intervention to have a measurement of the nucleus; Nature itself is making measurements with these vacuum fluctuations. Once in a while, one of these evanescent particles bumps into the nucleus, makes a measurement, and transmits that information into the environment. Since the nucleus is a pretty small target, this happens relatively rarely (in general), but even in a vacuum, even shielded from all external influences, a nucleus in a superposition of an undecayed and a decayed state is—at random times—measured by Nature. The superposition suddenly collapses, and the nucleus must "choose" whether to stay whole or break apart. To an outside observer, unaware of Nature's measurement, it looks like the nucleus simply breaks apart all of a sudden for no good reason. And because of this random choice that takes place during every measurement, it is impossible to say for certain when any given nucleus will decay: it is inherently a random event. While it's easy to say how an ensemble of these nuclei will decay, just as it's easy to say how a box full of gas will behave, it's as impossible to predict the behavior of a single nucleus as it is to predict the behavior of a single atom that careens randomly around the box.

This constant measurement is an unavoidable consequence of the rules of the quantum realm. It also is what holds the secret to the paradoxes of Schrödinger's cat. Here, then, lies the answer to one of the main questions of quantum theory: Why do microscopic objects behave differently from macroscopic ones? Why can atoms exist in superposition while cats can't? The answer is information. The transfer of quantum information into the environment—the constant measurement of objects by Nature—is what makes a cat different from an atom and the macroscopic different from the microscopic. Information is the reason that the laws of the quantum world don't seem to apply to large objects like baseballs and people.

As always with the Schrödinger's cat paradox, let's start small. Imagine that we have a quantum object, say, a large molecule like a seventy-carbon-atom fullerene. We can set it up in a state of superposition—

record the qubit **(0 & 1)** upon the molecule by passing it through an interferometer, forcing it to be in two places at once. How long can that qubit remain undisturbed? In theory, the molecule will stay in superposition as long as it remains unobserved—so long as no observer (Nature included) gathers information about the quantum state of the object. As long as the molecule is undisturbed, it can exist happily as a Schrödinger cat, neither here nor there but at both places at the same time. This is, in essence, what Anton Zeilinger's lab at the University of Vienna has done numerous times.

But it's not easy to keep the molecule undisturbed. If it is sitting out in the open air, molecules of nitrogen and oxygen are constantly bumping into it. When a nitrogen molecule slams into the fullerene, it makes a measurement; it gathers information about the fullerene. In fact, the nitrogen molecule and the fullerene become somewhat entangled.

After the collision, the nitrogen molecule carries information about the fullerene molecule. By looking at the ricochet, for example, you get information about where the molecule is. Therefore, if you make a measurement of the nitrogen's trajectory, you also get information about the fullerene molecule. And that is the essence of entanglement: gather information about one object and you automatically get information about another. So the fullerene and the nitrogen are entangled, thanks to that flow of information from one to the other. And when the nitrogen molecule collides into another air molecule, say, an oxygen, the oxygen "measures" the nitrogen and becomes entangled as well. If you had a sensitive enough particle tracker, you would be able to get information about the fullerene's position by measuring the track of the oxygen and working backward. Information about the fullerene is now resident on the nitrogen *and* the oxygen. And as these molecules collide with other molecules that collide with other molecules in the air, the information dissipates among all the molecules of the air; information about the fullerene spreads far and wide as the fullerene becomes entangled with its environment.

This flow of information from the fullerene to the environment makes it impossible to maintain the superposed state; the superposition collapses, and the fullerene "chooses" to be in state **(0)** or state **(1)**. This process of an object's gradual and increasing entanglement with its environment—the flow of information about an object into its surroundings—is known as *decoherence*.[8]

Decoherence, then, is the key to understanding how microscopic and macroscopic objects are different. When information flows from an object to its environment, it loses its superposition; it behaves more and more like a classical object. So, in theory, we would be able to keep a cat in superposition—we'd have a true alive-and-dead cat—if we could stop it from leaking information into its environment. We would have to stop decoherence.

How can we stop decoherence, even with a relatively small object like a fullerene? How can we stop information about the molecule from escaping away? The obvious way is to minimize the number of other molecules that bounce into our fullerene. For one thing, we should put it in a vacuum. That gets rid of the molecules of air that are ricocheting about the chamber; with a good vacuum, we can ensure that no molecules of air slam into the fullerene for the duration of the experiment. (And by chilling the chamber of air, we cause those molecules of air to slow down, reducing even further the probability that one slams into the fullerene.) We should also shield the fullerene from light—from photons that bounce off it—which also entangle with the fullerene and whatever else they scatter off of. But even in a perfectly dark room, absent of any particles, the fullerene can spontaneously announce its presence.

All objects radiate light. Any molecule that is not at absolute zero has a chance of emitting a photon—releasing a little dollop of energy

---

8. When it was first proposed, decoherence was so goofy sounding that it got one of its early proponents, Hans Dieter Zeh, in hot water. But it has since become completely mainstream, and it's been observed. In some ways, it is akin to entropy, but as will become clear later, it is arguably a more fundamental phenomenon.

into the environment in the form of light. This photon carries information about the object that it came from; it is automatically entangled with the object that radiated it and carries information out into the environment. It helps decohere the object, and there is nothing you can do to stop this process. But you can minimize it: the cooler the object is, the less energy it has to radiate away and the fewer photons it emits. So, the cooler the object is, in general, the slower decoherence happens. In February 2004, Zeilinger's lab published a paper that showed how increasing temperature increased fullerenes' decoherence rates. As the molecules got hotter and hotter, their interference fringes—the external sign of being in a superposed state—diminished and disappeared. So, in general, the cooler an object is, the longer it will stay in superposition.

Now what about an everyday macroscopic object like a cat? Imagine for the moment that we have put a cat in superposition—we've stored a **(0 & 1)** qubit on the cat. How long can that qubit remain on the cat?

Well, we are immediately in trouble. There is air surrounding the cat, so we have to put the cat in a vacuum to try to minimize the number of molecules that bounce off the cat and measure it. Even ignoring the (unpleasant!) effects that putting a cat in a vacuum chamber would have, it's a fairly impractical thing to do. Unlike the fullerene molecule, which presents a very small target for a molecule of air to ram into, the cat is a huge target. Even in a very, very good vacuum, there are thousands of molecules about. With a small object like a fullerene, this doesn't matter, because the probability of any given air molecule's slamming into it is extremely small; you need a lot of air molecules in the chamber to have a fighting chance of a collision with such a tiny target. But with a huge cat in a box, it's extremely likely that many collisions are occurring at any given moment even in a good vacuum: a big object is measured much more frequently than a small object is. The same goes for the measurements by photons and other particles in the environment; they are much more likely to hit a big cat than they are to hit a small fullerene. All of these measurements disseminate information about the cat into the environment.

Likewise, even if we chill the cat to near absolute zero, it will still emit quite a bit of radiation, at least compared to a microscopic object like a fullerene. Any given atom has a chance of radiating a photon at a low temperature. The lower the temperature, the lower the probability of emitting that photon. Since a fullerene has only 60 or 70 atoms, if the temperature is relatively low you can prevent all of those atoms from radiating. Just make the probability of emission one in a thousand or so for the time period you want to store your qubit and you've got better than a 90 percent chance that none of the atoms will radiate. A cat, on the other hand, has roughly a billion billion billion atoms. With a one-in-a-thousand chance or a one-in-a-million chance or a one-in-a-billion chance or even a one-in-a-billion-billion chance of an atomic emission, you would be guaranteed to have atoms on the cat emitting photons. There is essentially a 0 percent chance that none of the cat's atoms will radiate. The bigger the object is, the harder it is to keep from spilling its information through radiation.

So, all told, the smaller something is, the less complicated something is, and the colder something is, the less it decoheres. The bigger, sloppier, and warmer something is, the faster information about it leaks into the environment, despite the best efforts to isolate it. Scientists have calculated that in a perfect vacuum in deep space near absolute zero, even something so small as a dust particle a micron across—ten times smaller than the thickness of a human hair—would decohere in a millionth of a second. Store a qubit on it and Nature will make measurements and destroy the superposition in a tiny fraction of a second. If it is that hard for a tiny dust grain, imagine how hard it would be for something much warmer, sloppier, and larger—such as a cat in a box.

This is the essential difference between the microscopic, quantum world and the macroscopic, classical one. Nature has a harder time gathering information about cold, small objects, so they can preserve their quantum information for a relatively long time. But it's easy for Nature to gather information about big, warm objects, which pretty much describes everything that we encounter in day-to-day life. Even when quantum information does get inscribed upon a large object like a baseball or a cat, that information quickly disseminates into its environment, destroying whatever superposition it had. Large objects quickly become entangled with the environment as the information about the objects flows into the objects' surroundings.

Information—and decoherence—holds the answer to the paradox of Schrödinger's cat. When Schrödinger proposed his thought experiment, he got most of the details correct, but he was unaware of the effects of decoherence. Yes, a particle can be in a superposition. Yes, you can transfer that superposition, that qubit, from the particle to the cat. Yes, the cat can be put in a superposition of some sort, at least in theory. But because the cat is big and warm, the information about the cat's state leaks into the environment even before someone opens the box. The cat's state decoheres in a tiny, tiny, tiny fraction of a second. The cat's superposition disappears in such a small time that it is not noticeable at all; effectively, it instantly "chooses" to live or die. Even though the cat is following the laws of quantum mechanics, it behaves like a classical object; you're never going to be able to catch the cat in a state of superposition or make a cat-interference pattern. The flow of information into the environment is too fast. Nature measures the cat long before anyone can open the box. Even in a completely isolated environment, Nature has the power to make measurements, and large, warm objects are more easily measurable than small, cold ones.

Decoherence is what kills the cat. And decoherence is what makes a macroscopic object behave classically while a microscopic one shows quantum behavior. Including our brains.

Brains are information-processing machines and are subject to the laws of information. Classical information theory seems to imply that we are merely extremely complex information-processing machines. This would mean that we are not fundamentally different from a Turing machine or a computer. Obviously, this is a disturbing conclusion, but there is one obvious way out. If the information in our heads is *quantum* information rather than classical information, then our minds take on a whole new dimension.

To some investigators, the phenomenon of quantum superposition and collapse seems strikingly similar to what goes on in the mind. In the quantum realm, Schrödinger's cat is neither alive nor dead until some process—measurement or decoherence—leaks the information into the environment, collapses the superposition, and forces the cat to "choose" life or death. Similarly, the human mind seems to grasp at multiple, half-formed ideas, all flitting below the threshold of awareness at the same time. Then, somehow, something snaps—an idea solidifies and winds up at the front of consciousness. Ideas start out in superposition in the preconscious and then wind up in the conscious mind as the superposition ends and the wave function collapses.

Quantum consciousness aficionados suspect that the analogy might be more than a coincidence. In 1989, the British mathematician and quantum theorist Roger Penrose publicly joined their number, speculating in a popular book called *The Emperor's New Mind* that the brain might be acting like a quantum computer rather than a classical one. But neurons, as we saw above, tend to behave just like classical devices that store and manipulate bits; if the brain is somehow storing and manipulating qubits, there must be another mechanism besides the neuron's standard chemical bit flip that biologists are familiar with.

The anesthesiologist Stuart Hameroff of the University of Arizona was interested in consciousness for a different reason from that of philosophers: he was trained to remove it and restore it. Even as advanced as anesthesiology has become, medicine has a very primitive and unsatisfying understanding of the phenomenon of consciousness; there isn't even a good definition for it. So it is an active source of study, and Hameroff found himself attracted to it. During his studies of neurophysiology, trying to understand consciousness, Hameroff came across a possible seat for quantum nature in the brain: *microtubules*, tiny tubes constructed out of a protein called tubulin. These tubules are structural; they make up the skeletons of our cells, including neurons. But what makes them interesting is not their classical role, but their (potentially) quantum one.

Tubulin proteins can take at least two different shapes—extended and contracted—and since they are relatively small, in theory they might behave like quantum objects. They might be able to take both states, extended and contracted, at once in a state of superposition. The tubulin might be able to store a qubit. It is also possible that an individual tubulin protein might affect its neighbors' quantum states, which in turn affect their neighbors', and so forth, throughout the brain. In the 1990s, Penrose and Hameroff showed how such a tubulin-based quantum messaging system could act like a huge quantum computer. And if we've got a quantum computer running in parallel to the traditional, classical computer of the brain, the quantum computer might be where our consciousness resides. It would explain why we are more than mere calculating machines—we would be quantum, not classical.

This idea of the quantum brain attracted a few physicists, some consciousness researchers, and a large number of mystics. Most neurobiologists and cognitive scientists, however, didn't put much stock in the idea. Neither did quantum physicists; it was too speculative. Besides, the brain is a terrible place to do quantum computing.

Quantum information, by its nature, is very fragile. Nature is constantly performing measurements and dissipating stored qubits, entangling them with the environment. Qubits tend to survive best when they are stored on a small object, isolated in a vacuum, and kept very cold. Tubulin proteins are fairly large compared to quantum objects like atoms, small molecules, and even larger molecules like fullerenes. Worse yet, the brain is warm and (usually) much more full of stuff than a vacuum is. All these things conspire to dissipate quantum information that might be stored on a tubulin molecule. In 2000, Max

Tegmark, a physicist at the University of Pennsylvania, plugged in the numbers and found out just how bad an environment the brain would be for quantum computation.

Combining data about the brain's temperature, the sizes of various proposed quantum objects, and disturbances caused by such things as nearby ions, Tegmark calculated how long microtubules and other potential quantum objects within the brain might remain in superposition before they decohered. His answer: the superpositions disappear in 10^{-13} to 10^{-20} seconds. Because the fastest neurons tend to operate on a timescale of 10^{-3} seconds or so, Tegmark concludes that whatever the brain's quantum nature is, it decoheres far too rapidly for the neurons to take advantage of it. While many quantum consciousness aficionados still argue that the mind has a quantum nature, it is hard to think of a way that it could be so: decoherence is too powerful a phenomenon. The brain appears to be classical after all.

Even if the human brain is "merely" a machine for manipulating and storing information, it is so complex and intricate that scientists have no real idea about how it does what it does except in a very gross manner. Philosophers and scientists have a hard time even defining what consciousness is, much less understanding where it comes from. Is consciousness something that simply emerges from a sufficiently complex collection of bits moving about? Scientists have no compelling reason—other than squeamishness about what it means to be human—to say it isn't. Even if our brains are nothing more than very complex information-processing machines, they operate on a different level and timescale than the information processors in our cells. However, like our genes, our brains are following the laws of information—and decoherence. Information theory sees no fundamental difference between the brain and a computer, just as it sees no fundamental difference between the macroscopic world and the microscopic. Decoherence shows that our brains cannot be quantum computers, just as it explains why Schrödinger's cat can't behave the way an atom does—and why subatomic particles behave so differently from macroscopic ones.

Decoherence is not a complete answer to what makes quantum mechanics so weird, but it is a large step toward understanding the nature of the quantum universe and showing that you don't need separate laws to describe the quantum world and the classical world. The quantum laws hold just fine on all scales, and it is Nature's constant gathering and dissemination of information that makes microscopic and macroscopic objects display such different behavior.

Throughout this process, Nature is manipulating information. It measures it, transfers it, rearranges it. But as far as scientists can tell, Nature never destroys information or creates it. Decoherence is not a matter of getting rid of information; when a superposition collapses and a qubit on an object such as an atom gets "wiped out," it is transferred to the environment, not destroyed. Indeed, the process of decoherence obeys two laws of quantum information known as the no-cloning and no-deletion rules. These rules, which follow from the mathematics of quantum theory, state that qubits can be moved from place to place but can never be duplicated with perfect fidelity or be entirely erased. Thus, decoherence is neither creating information nor destroying it. Nature is just taking information from an object and spreading it out into the environment. Information seems to be conserved.

This dissemination of information—decoherence—is analogous to something we have already seen. If we place atoms of gas in the corner of a container, they will quickly spread out to fill the entire container; the entropy of the system will quickly increase. (Also, if we cool down the container, the atoms move slower and spread out less quickly.) Even though it is purely a statistical phenomenon, it is as if Nature is conspiring to spread the atoms around. Similarly, if we put information on an object, the random motion of particles and the fluctuations of the vacuum conspire to spread that information about, dissipating it into the environment. Though the information still exists, it gets harder and harder to retrieve as the process of dissipation continues. Like entropy, decoherence is a one-way phenomenon: though it is *possible*, it is astronomically improbable that Nature will gather information from the environment and deposit it on a macroscopic object, putting it in superposition. Like entropy, decoherence lets you know which way time is running; decoherence is an arrow of time. And the two are linked. Decoherence of a qubit increases the entropy of the system by, you got it, $k \log 2$.

In some ways, decoherence is even more fundamental than entropy. While the entropy of a gas in a container increases on time scales of microseconds, decoherence operates on timescales billions and billions of times shorter. Entropy only tends to increase when a system is out of equilibrium, while Nature is always measuring and disseminating information; thus decoherence occurs even when a system is in equilibrium. And while the concept of entropy leads to the second law of thermodynamics, the idea of decoherence is related to what may be an even more powerful law, a new law:

INFORMATION CAN BE NEITHER CREATED NOR DESTROYED.

This is a law that encapsulates the laws of thermodynamics and the explanation for the weirdness of quantum mechanics and relativity. It describes the way physical objects interact with one another and the way scientists can acquire understanding about the natural world. It is the new law.

But the new law, the law of information, has not yet been firmly established. Though many scientists believe that it holds, a number of challenges and potential exceptions to the law have yet to be resolved. The most serious challenges come from the theory of relativity, for the laws of Einstein and the law of information seem to be at odds.

Nevertheless, when the theory of relativity faces off against the law of information, the law of information seems to win. Information may even be able to survive what nothing else in the universe can: a trip straight down the maw of the most destructive power in the cosmos—a black hole.

CHAPTER 8

CONFLICT

Awe very properly hangs about it, since it is the immovable standard
and silent witness of all our memories and assertions; and the past and
the future, which in our anxious life are so differently interesting and
so differently dark, are one seamless garment for the truth, shining like
the sun.

—George Santayana

Information is at the heart of the mysteries of quantum theory, just
as it is responsible for the paradoxes of relativity. But scientists don't
have a complete theory of quantum information yet, so they don't
know the answers to all the difficult philosophical problems that these
theories raise. Though decoherence seems to explain the apparent difference between the microscopic and the macroscopic as well as the
paradox of Schrödinger's cat, a lot of questions are still unresolved.
The most serious ones have to do with relativity.

Physicists still do not understand the mechanism for entanglement. They are forced to accept that particles somehow "conspire"
across vast distances. The laws of quantum theory and quantum
information describe entanglement wonderfully. However, they do
not explain *how* entanglement works; they don't reveal how entangled
particles manage to conspire. To find out, scientists are pushing into
stranger and stranger territory: they are exploring a realm so far out
that some of the best quantum experiments are brushing up against
the realm of the paranormal—telepathy. Einstein's "spooky action" is

spookier and seemingly more unrealistic than a ghost story, and it is making scientists question their notions about the flow of time.

Spookier still is the mystery of the black hole. These collapsed stars devour anything that comes within their grasp—including light—and everything that crosses an invisible threshold is irreversibly destroyed. Or is it?

If information is truly conserved, if it can be neither created nor destroyed, not even a black hole can eradicate the information that it devours. Black holes might be enormous information storage devices, keeping quantum information intact and disgorging it many billions of years later. In fact, until 2004 the most famous scientific wager in the world, a bet between Stephen Hawking and Kip Thorne on one side, and John Preskill on the other, had to do with whether information is destroyed or preserved when it falls into a black hole. While this might seem like a frivolous bet, it goes directly to the core of what laws Nature truly follows. If information is conserved, then it can penetrate where no telescope, no robotic probe, no observer can ever go. Information will give us a way to peer behind the shroud that protects the black hole from prying eyes. Information will reveal the secrets of the most mysterious objects in the universe, regions where the laws of physics break down and quantum theory and relativity are most directly in conflict.

Understand information and you understand black holes. Understand black holes and you understand the ultimate laws of the universe. It was a wager with very high stakes indeed, and when the wager was settled, it created headlines all across the world.

The conflict between relativity and quantum mechanics still shakes physics to its roots, and scientists across the world are trying to figure out the consequences of that conflict. For example, the idea of entanglement threatens to undermine the speed-of-light limit of information transfer at the heart of relativity: If particles were conspiring at great distances to come up with equal and opposite quantum states after a measurement, then could they be used to send a message faster than the speed of light? Theorists say no, as will soon become clear. But that doesn't stop some from hoping that entanglement holds the secret to a new form of communication. Among these is Marcel Odier, a Swiss philanthropist who made his millions in banking. Odier and his wife, Monique, created a foundation to explore a realm they have dubbed "psycho-physics"—the half-scientific, half-occult realm where physics and parapsychology cross.

Odier is convinced that people—and animals—can be telepathic, and his foundation has funded a number of studies to explore this phenomenon. While he says he has sufficient evidence to believe in telepathy, he doesn't know what mechanism would allow people's minds to connect with one another. However, quantum mechanics seemed to provide a way: entanglement. Odier hoped that entanglement could help explain the mechanism of telepathy, so he spent about $60,000 to fund an experiment at the University of Geneva: Nicolas Gisin's attempt to figure out the "speed" of quantum entanglement.

Gisin—like most serious scientists—thinks that telepathy is bunk. Nevertheless, the phenomenon of entanglement is so weird that it attracts the attention of fans of the paranormal, including Odier. And Gisin had no problem accepting the money; it allowed him and his colleagues to perform a first-rate experiment. Though Gisin's group didn't find any clues to a mechanism for transmitting information from person to person via entangled particles—indeed, as will become apparent, the laws of quantum information show that it is impossible to send messages with entanglement alone—Gisin found something almost as disturbing as telepathy. His experiments proved that there was a fundamental conflict between the theories of relativity and quantum mechanics about the nature of time. In quantum theory, unlike in relativity theory and in everyday life, there may be no such thing as "before" and "after."

Theorists have long known that relativity and quantum theory are at odds. Relativity is a smooth theory. It deals with the nature of space, time, and gravity and treats the fabric of space and time as a smooth, continuous sheet. Quantum theory is a rough, grainy theory. It deals with packets and quantum jumps, chunks of energy and a discrete, broken-up view of the universe. Relativity and quantum theory are very different ways of picturing the universe, very different mathematical methods that often don't agree. Most of the time they don't come into direct conflict; relativity tends to deal with galaxies and stars and things moving near the speed of light—the realm of the very large and the very fast. Quantum mechanics becomes important for atoms and electrons, neutrons and other tiny particles—the realm of the very small and, often, the very cold and slow moving. These are very different regimes, and most of the time they don't overlap. Most of the time.

Entanglement is one of the areas where the two theories go head to head. Einstein put a speed limit on the transmission of information, yet quantum theory says that entangled particles *instantly* feel when their partners are measured. Quantum theory is agnostic about how the particles conspire with each other, while Einstein's theory is very, very careful about defining how messages get sent from place to place. This is a key source of conflict, and it is precisely the region that Gisin was trying to understand.

In chapter 6, I described how, in 2000, Gisin created sets of entangled photons that sped in opposite directions in fiber-optic cables around Lake Geneva; when he measured one, the other instantly felt the measurement. If the two particles were somehow sending a message to one another, then that message would have had to travel at more than ten million times the speed of light to get from one to the other in time to effect a successful conspiracy. As it happens, the measurement of this "speed of entanglement" was incidental. In the 2000 experiment, funded by Marcel Odier, and in a 2002 follow-up experiment, which was not, Gisin tried to force Nature to reveal the nature of the entanglement conspiracy. He tried to ruin the particles' entanglement in an Einsteinian way, and when he failed, he showed that the concepts of "before" and "after" don't apply to quantum objects in the simple way they do to relativistic ones.

Gisin's trick was to make the entangled pair act out the spear-and-the-barn paradox. As described in chapter 5, the paradox uses the relative motion of two participants to make them disagree on the order of events. Observer A (the stationary bystander) thinks that the front door of the barn closes before the back door opens; observer B (the sprinter) thinks that the back door opens before the front door closes. So long as the front door and the back door are not causally connected, both observers can be correct at the same time even though they disagree about the order of events.

In their Geneva lab, Gisin and his colleagues sent sets of entangled, superposed photons toward the villages of Bernex and Bellevue in a classic EPR experiment. But there was a twist: their laboratory setup was in motion. In the first experiment, they had a detector rotating very rapidly, making it act like the runner in the spear-and-the-barn paradox.

Thanks to its motion, from the moving detector's point of view it measured particle A before particle B struck the detector in the other village. As soon as particle A struck the moving detector, its superposition collapsed because of the measurement. If there is some form of "communication" between the two particles, B must learn about A's collapse and somehow collapse as well. A's superposition collapsed because of its *own* measurement while B's collapsed because of its *partner's* measurement.

But from the stationary detector's point of view, the situation was reversed. In the stationary detector's frame of reference, particle B struck the detector and was measured before particle A reached the moving detector. From the stationary detector's point of view, particle B's superposition collapsed because of its own measurement while A's collapsed because of its partner's.

If the measurement of one particle somehow affects the other—if some sort of communication between the two particles allows them to conspire—Gisin's experiment showed that it is impossible to tell which is the *affecter* and which is the *affected,* which is the sender of the message and which is the receiver. This is a ridiculous state of affairs; if there's any form of communication between one particle and the other, then the disagreement about which particle was measured first means that they disagree about which particle initiated the conspiracy and which merely followed.

The 2002 experiment was a refinement of the first one. It used moving beam splitters instead of a moving detector. As with the first experiment, this setup produced the same result: the detectors disagreed on which particle struck first.

If there is some sort of message going from particle to particle, there is no well-defined sender or well-defined receiver. The particles seem to ignore the concepts of before and after. They don't care which was measured first or last, which was the sender or the receiver. No matter how you set up the experiment, the entanglement remains unhindered; the two particles conspire to end up in opposite quantum states even though neither "chooses" its state until the act of measurement forces it to. Telepathy is nonsense, but the quantum world is even stranger than a parapsychologist's fantasies.

Gisin's experiment was a dramatic example of the difficulty of describing entanglement within a framework of exchanging a message. It is natural to think that the two particles must somehow communicate with each other; on the face of it, there seems to be little alternative. Scientists have proved that the particles' superpositions don't collapse until the act of measurement or decoherence; the particles can remain in an ambiguous amalgam of two states so long as they remain undisturbed. When one of the particles is measured, though, *both* superpositions collapse, and the collapses always are correlated with each other: if one decides to be spin up, the other will be spin down; if one is horizontally polarized, the other will be vertically polarized. The collapse of the wave functions happens at the same time and in a correlated way, yet the collapse is inherently a random event that cannot be decided ahead of time. The only obvious way out of this seeming contradiction is to assume that the two entangled particles are somehow communicating. But Gisin showed that this communication, if there is such a thing, is a very odd sort of message indeed. It moves much faster than the speed of light, and it doesn't matter which is the sender and which is the receiver, yet the message gets through all the same.[1]

In fact, it's best not to think about entanglement as an exchange of a message, because a message implies that information is being sent from one of the particles to another. And it has long since been established that one of a pair of entangled particles cannot transfer information to the other through its "spooky" influence. In the 1970s, the physicist Philippe Eberhard proved this, mathematically. It is impossible to use an EPR pair to transmit information faster than light—and Gisin's experiment is a good demonstration of why this is. Even though the quantum states of particle A and particle B are correlated—the quantum state of one depends on the quantum state of the other—there isn't a *causal* relationship between the two. The measurement of particle A doesn't really signal its twin to "collapse now"; A isn't "causing" B's collapse any more than B is "causing" A's collapse. They just happen to collapse simultaneously and don't care a fig about which was measured first or Einstein's concept of causality. There is no good explanation for why this is; it just is. It is a consequence of the mathematics of quantum theory, but it doesn't have a good, intuitive, physical reason behind it.[2]

This is a very weird state of affairs, but it is one that physicists have come to accept. Nobody has been able to use the spooky action of EPR

1. Even more bizarre, entangled particles will, at least in theory, show this correlation even if the measurements occur *before* the particles are entangled with each other. This is known as a "delayed choice" experiment, and it means that the state of entanglement exists even before the particles know they are entangled.
2. Though there are some promising leads—more on this in chapter 9.

pairs, even in theory, to send a bit, a **0** or a **1**, or a qubit like **(0 & 1)** from place to place faster than light. This is despite the fact that physicists are able to *teleport* an object across a laboratory using entanglement.

The term *teleportation* is misleading, but that's what this process's inventor, the IBM physicist Charles Bennett, chose. The word teleportation conjures *Star Trek* visions of disassembling Mr. Spock in a flash of light and then reassembling him on the planet's surface. Quantum teleportation is very different from that. It teleports information, not matter.[3]

In 1997, two teams of physicists, led by Francesco De Martini at the University of Rome and Anton Zeilinger at the University of Vienna, successfully used an EPR pair to transmit a qubit from one atom to another. The details of the experiments were slightly different, but the essence was the same. They simultaneously measured one of a member of the EPR pair along with a particle that stored a qubit, entangling the two. On the other end of the lab, they measured the other member of the EPR pair along with an empty, target particle that would receive the qubit. This sets up a chain of entanglement: the qubit-storing atom is entangled to an EPR particle, which is entangled to the other EPR particle, which is entangled to the target atom. A few manipulations later and the qubit is transferred from the source atom to the target atom. Owing to the no-cloning rule, the original copy is destroyed, but the quantum state of the atom is transferred across the lab on the back of the spooky action at a distance.

---

3. The difference might be moot. Quantum mechanics doesn't distinguish between particles; one electron is identical to every other electron in the universe, for example. The only difference is in their quantum state—the quantum information that they carry. If you take the quantum state of electron A and transmit it across the universe and reconstruct it on electron B, then there is no distinction between the original electron (whose quantum state is now destroyed, due to the no-cloning rule) and the one that's been reconstructed at the end of the teleportation process. In a sense, Mr. Spock wouldn't really survive the teleportation process. He's destroyed while an exact duplicate steps out of the other transporter. But if nobody can tell the difference between the original Spock and the duplicate—not even the duplicate himself—is he really a copy or is he the original? That's a question for philosophers, not scientists, but I must admit that I'd refuse to go on a *Star Trek*–type teleporter if one did exist.

If you are using EPR pairs to transmit a quantum bit of information, aren't you violating the ban on instantaneous information transfer? No, because the teleportation process has one catch. It needs classical information to be transferred from the sender to the receiver as well, two classical bits that can be transmitted at best at the speed of light. The "few manipulations" can't be performed without those two bits of information; without the classical bits, there is no way to know how to reconstruct the qubit on the target particle. Though spooky action at a distance is quantum teleportation's mechanism for transmitting the quantum state of an atom onto another, the actual information on the atom can only travel from place to place at the speed of light. There is no way to violate the ban on sending information faster than light speed.

Einstein's ban on transmitting information faster than the speed of light holds fast, despite the weirdness of the spooky action in entanglement. Entanglement does not derail the laws that dictate how information behaves. However, entanglement still extracts a great cost. Quantum states collapse instantly, ignoring Einstein's careful stress on the concepts of before, after, and causality—and entanglement's mysterious conspiracy remains as dark as ever.

Scientists don't yet truly understand entanglement, but the laws of quantum information appear to be safe from the threat. However, another dark mystery threatens to derail the concept of the conservation of information—the darkest objects in the universe. Black holes.

A black hole is the nightmare legacy of Einstein's theory of relativity. It is a gaping wound in the fabric of spacetime, an unfillable hole that gets bigger and bigger as it swallows matter. It is shrouded by a curtain that shields it from prying eyes—even Nature's—as no information passes from the center of a black hole to the environment outside. Indeed, the region near a black hole is cut off from the rest of the cosmos. In a sense, each black hole is its own universe.

Black holes are massive stars that have died a spectacular death.[4] Throughout a star's lifetime, it is a cloud of (mostly) hydrogen in a state of tenuous equilibrium. On one hand, the sheer mass of the star—the gravitational force it exerts on itself—tries to shrink it down to a point. On the other hand, the nuclear reactions that rumble in the star's furnace, where the star converts hydrogen to helium and to heavier elements, try to blow it apart. For millions or billions of years (depending on the star's mass), these two forces balance out each other; gravity is unable to crush the star because of the outward force of the fusion reaction, while the fusion furnace at the center of the star cannot blow the star apart because the star's matter is held in by gravity.

But when a star begins to run out of fuel, that balance is upset. The fusion furnace sputters and flares as it uses up different sorts of fuel. The star shrinks and bloats and shrinks again. At some point, the star's fuel is exhausted. The outward force ceases and the only force left is gravity, unchecked by the force of fusion. A sufficiently massive star collapses rapidly into itself, creating an enormous explosion: a supernova, the most violent event in the universe.

Much of the star's mass is blown away in a violent burst of energy, but a good proportion of it is held, trapped by the gravity of the collapsing star—which gets smaller and smaller and smaller in a tiny fraction of a second. If the star is large enough, the force of gravity is so strong that nothing can stop its collapse; it gets denser and denser as it gets tinier and tinier. It gets smaller than our sun, smaller than the Earth, smaller than the moon, smaller than a basketball, smaller than a grapefruit, smaller than a pea, smaller than an atom. As far as scientists know, nothing in the universe can stop the star from shrinking

4. This is true of ordinary black holes, those that are only tens or hundreds of times more massive than our sun. There are other classes of black holes, such as the supermassive black holes that sit in the center of galaxies. The one at the heart of our galaxy, Sgr A* (Sagittarius A*), weighs about as much as 2.5 million suns, and scientists are less certain how it formed, though the same physical principles apply to supermassive (and intermediate-size) black holes as to the run-of-the-mill variety.

into nothingness; the mass of tens or hundreds of suns is packed down into no space at all. It has become a singularity—a point of infinite density, where the curvature of space and time becomes unbounded. The black hole is a bottomless pit in spacetime, an infinite tear where time and space no longer truly have meaning. And because of this— because it is a very massive object, subject to the laws of relativity as well as being a very tiny object subject to the laws of quantum mechanics—black holes are regions where the two theories come into direct conflict. By studying this tear in spacetime, the singularity at the heart of a black hole, scientists would quite possibly be able to resolve the conflict between the two theories. The result would be a single, unified theory that holds on all scales and in all regions of the universe. It would be the ultimate achievement of physics.

Unfortunately, studying a black hole is out of the question, even in theory. The wound in the fabric of the universe is not an open wound. The singularity of the black hole is surrounded by a shield that protects it from prying eyes. Though this shield is not a physical object— you would not notice it if you were to pass through it—it marks the boundary between two universes. Anything that crosses this *event horizon* will never escape the clutches of the black hole; not even light can move fast enough to propel itself away from the gravitational pull of the collapsed star.

Black holes were so named by the Princeton physicist John Wheeler, who realized that such a monstrosity would be the darkest object in the universe. Because the massive star absorbs whatever light and matter that happen to cross its one-way barrier, it would appear as a large dark blotch in the heavens.

Scientists are a decade or more away from being able to view the blackness of a black hole directly; at the moment, they are only able to infer a black hole's presence by the motion of stars around it. At the center of our galaxy, for instance, massive stars wheel around an enormous, invisible mass that is as heavy as millions of our suns. The stars' motion is caused by the gravitational attraction of a black hole. Even though the black hole is invisible, scientists can see how it pulls on stars and wolfs down matter.

But even with the most powerful telescopes in the universe, seeing the silhouette of a black hole would not tell us about the singularity, the tear in spacetime at the collapsed star's heart. In fact, even if we were able to dump a probe into the maw of the black hole, the probe would be unable to tell us anything about the singularity or the region hidden by the event horizon.

Imagine that we are in a research spaceship orbiting a safe distance from a black hole. The vessel is equipped with a disposable probe—a little robot that sends a coded message back to the mother ship every second. Beep. Beep. Beep. That probe is built really tough, tough enough to withstand the gravitational forces that will try to tear it apart and the radiation that will try to fry its circuits; no matter how the black hole tries to destroy it, that probe will emit one message, one beep, every second until the end of time.

Now let's fire the probe out of the ship toward the black hole. From the probe's point of view, it emits a chirp once a second, every second, as it flies toward the collapsed star. It observes lots of strange visual effects owing to the gravitational bending of light; all the stars in the universe seem to crush together, eventually filling up less than half the sky. But the probe continues on, chirping merrily along. Crossing the event horizon isn't much of an event after all; it radios back, "I'm about to cross the event horizon . . . now!" when it traverses the barrier, but it doesn't see any physical barrier or anything to indicate that it has crossed into the realm of no return. Nothing unusual happens. It keeps on beeping and beeping and beeping, once a second as it falls toward the singularity. These beeps contain information about what the probe is seeing; having crossed the event horizon, it is radioing information about the realm behind the curtain that shields the black hole. The probe will fall into the singularity, into the center of the black hole, and disappear—beeping away until the very end. Our probe has sent us valuable information about the unknown region near the heart of a black hole.

The only problem is that these invaluable messages never reach the mother ship. Even though, from the probe's point of view, it sent a message every second, Einstein's theory of relativity tells us that gravitational fields affect space and time just as does rapid motion. So, from our mother ship's point of view, the probe's clock gets messed up as it approaches the black hole. It slows down. The beeps get farther and farther apart as the probe gets closer and closer to the black hole: 1.1 seconds apart, 1.5 seconds apart, 3 seconds apart, 10 seconds apart, 2 minutes apart, and on and on. As the probe approaches the event horizon, the messages get sparser and sparser. Stranger still, it gets harder and harder to see the probe. The light coming from the probe gets redder and redder and dimmer and dimmer as the probe approaches the event horizon. Soon, it is invisible to human eyes, and even a sensitive infrared telescope aboard the ship would be having trouble spotting the probe, which appears to still be falling toward the event horizon.

We keep observing the probe and recording the ever less frequent messages that come days apart. Weeks apart. Years apart. Decades apart. After years and years and years of observation, the probe is only an incredibly faint shadow hovering near the edge of the event horizon—but never crossing. Eventually, we get a drawn-out message from the probe. It takes several years to receive the signal: "Iiiiii'mmmm aaaaaboooooouuuut toooooo crrroooosssss theeeeee eeeeeveeeennnnntttt hooooorrrrriiiizzzzooooonnnn . . . ," but the concluding word of the message, "now!" never arrives. We have heard our last from the probe, which fades from view, hovering eternally on the edge of the event horizon—but never crossing.

No matter how we try, no matter how advanced our probes or our telescopes are, it's impossible to get any information from beyond the event horizon at all. Just as the intense gravity prevents any light from crossing the horizon and escaping the black hole, it prevents any information from doing so, too. This is the amazing property of an event horizon; it isolates the inside of a black hole from the rest of the universe; it blocks information from escaping. You can find out about the region behind the event horizon for yourself—just jump into the black hole—but you will never be able to share your discovery with anyone beyond the event horizon. You might discover just what lies at the center of the black hole—you might unravel the mystery of the singularity—but you will never be able to tell the tale to scientists back home, even with the most powerful transmitter in the universe. The event horizon is a cosmic censor; it prevents observers from learning what lies beyond it.

This information barricade is so complete that an outside observer can only get an extremely limited amount of information about the black hole. You can learn how big it is: by watching the way it affects nearby objects, you can figure out how much mass it has. You can figure out how fast it's spinning—how much angular momentum it has. (A spinning black hole has a somewhat flattened, oblate event horizon, and nearby objects are affected by the spin of the black hole in various subtle ways.) You can measure how much electric charge the black hole carries, though there is no reason to believe that black holes, in nature, carry a significant amount of charge. Other than that, black holes are pretty much a cipher. In a sense, they are the simplest objects in the universe because they are totally indistinguishable except for these three properties.

You can't tell what the black hole was made of. It could have been formed from a cloud of hydrogen gas, or an enormous brick of antimatter, or a clump of neutrons, even a big pile of Ford Pintos, for that matter. It is irrelevant what kind of mass, what material, went into the building of the black hole; all the information about (and stored upon) the matter is unattainable because that matter has disappeared behind the curtain of the event horizon. It is inaccessible, so we will never be able to tell whether all black holes formed from collapsing stars or whether there is an artificial one that has been made out of a critical mass of alien garbage bins. We can't tell what kind of mass went into the making of the black hole; all that we can distinguish is the *amount* of mass in a black hole and how it is spinning.

In the 1960s, Wheeler coined a phrase that summed up the nearly complete lack of information about a black hole's composition: "A black hole has no hair." That is, black holes have no distinguishing features: nothing sticks out beyond the event horizon to let you know what the black hole was made of.[5] The no-hair theorem is now one of the basic tenets of black hole theory; it was proved in the 1970s by Stephen Hawking and number of other physicists. The black hole swallows all information about its origins when it retreats behind its event horizon.

Not even Nature itself can gather information about the region behind the event horizon. All of Nature's probes, all of its measurement devices, are unable to penetrate the event horizon and return. Cosmic rays disappear down the maw of the black hole, as do the photons that suffuse the universe. Even the particles created by vacuum fluctuations get swallowed. There is nothing, nothing at all, that Nature or any other observer can do to retrieve information that has disappeared behind the event horizon. The information about the black hole's origin is lost forever to the cosmos.

This is a very, very troubling situation to an information theorist. In the last chapter, it seemed that information was always conserved. Nature could neither create nor destroy quantum information; it could rearrange it, store it, and dissipate it, but Nature could not extinguish information. But a black hole seems to be doing just that. Store a qubit on an atom and dump it into a black hole and that qubit is lost to this universe; indeed, all quantum information about that atom—including its very "atomness"—is gone. All that remains is the signature of the atom's mass, angular momentum, and charge, which have

---

5. According to the theorist Kip Thorne, the no-hair terminology, when translated into French or Russian, turned into an absolutely filthy phrase. One Russian editor even refused to publish a paper about the no-hair theorem because it was so obscene.

all been added to the black hole. Even Nature itself cannot divine whether we threw in an atom or a neutron or antimatter, much less what quantum state the atom was in and what quantum information it contained. That looks a whole lot like the destruction of information, and it would blow the new law, the law of conservation of information, to pieces. This is the information paradox of black holes.

You might have seen popular accounts of the black hole information paradox—usually containing some mumbling about tossing encyclopedias into black holes—but the articles seldom make much sense. That is because the problem is much deeper than the disappearance of the classical information that the encyclopedia contains. The paradox hinges upon the loss, at least to Nature, of all the quantum information about whatever lump of matter you toss in the black hole. Although there are very strong reasons for believing that information is conserved, the information in that lump of matter is gone. The information is inaccessible. But has it been *destroyed*? Is this information erased without any trace?

Nobody knows. But there's reason to believe that it isn't—that information survives even the ultimate torture of falling into a black hole.

It is impossible to retrieve information about the region shrouded by the event horizon, but that doesn't stop Nature from trying. It constantly probes with cosmic rays, photons, and vacuum fluctuations. And though these attempts don't retrieve any information, they *do* have a measurable effect.

Nature's last-ditch measurement scheme uses vacuum fluctuations, those particles that are constantly winking in and out of existence at every point in space. These particles tend to get produced in pairs—a particle along with its antiparticle—which are spontaneously born, fly away from each other for a moment, then crash back together, annihilating each other. But along the event horizon of a black hole, this state of affairs changes slightly. At the very edge of a black hole's event horizon, Nature creates particle-antiparticle pairs as always, but some of the time one of the particles crosses the event horizon and is trapped, while the other escapes, flying off into space. This particle contains no information about the interior of the black hole.[6] However, it—and billions of other particles born in the same way—owes its existence to its creation alongside the event horizon. An observer near the black hole would see the event horizon "radiating" zillions of these particles; even though a black hole swallows everything incautious enough to cross the event horizon, it still radiates matter and energy in the form of these particles that have lost their siblings. Nature's measurements, the fluctuations of the vacuum, cause black holes to radiate particles into space.

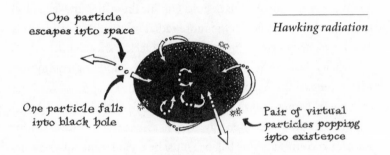

One particle
escapes into space

Hawking radiation

One particle falls
into black hole

Pair of virtual
particles popping
into existence

In the 1970s, Stephen Hawking proved that this radiation was as featureless as can be; it followed the so-called blackbody spectrum. In the nineteenth century, Ludwig Boltzmann and other scientists figured out how to describe the amount of radiation that streams from an idealized, featureless object—a blackbody—at a given temperature. Black holes behave like blackbodies, so the amount of radiation coming off them yields their "temperature." Black holes are very cold blackbodies, as the radiation they emit, Hawking radiation, is quite

6. Yes, the particle and its twin are entangled, yet remember, the mere spooky action at a distance isn't able to convey information. You've got to have a classical bit go from one to the other as well; you need to *compare* the measurements of one to the other if you are to extract any information. This, of course, is impossible as one of the pair has fallen past the event horizon. Therefore, even entangled particles yield no information about what the black hole is hiding.

feeble, but they are blackbodies with a finite temperature nonetheless. A black hole's Hawking radiation, whose properties depend on the curvature and size of the event horizon, reveals how hot the black hole is. Though the temperature isn't an extra piece of information—it can be inferred by the black hole's mass, spin, and charge—it shows that a black hole has a well-defined temperature and therefore can be analyzed with the laws of thermodynamics. It also carries the seeds of the black hole's demise.

Black hole thermodynamics is an odd-sounding field of study; black holes aren't containers of gas or ordinary chunks of matter. But the laws of thermodynamics are yielding some surprising insights about the properties of a black hole. For one thing, the smaller a black hole is, the hotter it is and the more radiation it emits per unit area. This has a curious consequence. It makes black holes explode.

A black hole with a finite temperature is radiating energy, and when something—even a black hole—radiates energy, it must get that energy from somewhere. (The particles from the vacuum fluctuations don't provide any energy; they're essentially "borrowed" from Nature's accounts and the balance must be repaid somehow.) A spinning black hole can use the energy stored in its rotation, slowing down as it radiates, but once it stops spinning, that source is gone. It must get that energy from somewhere else—and that somewhere else is the mass of the black hole itself. The black hole consumes its own mass to create the radiation. But a less massive black hole has a smaller event horizon; the event horizon shrinks and becomes slightly more curved. And the smaller the event horizon, the hotter the black hole gets: it emits more radiation. It shrinks some more and heats up again, emitting even more radiation. Smaller. Hotter. Smaller. Hotter. Faster and faster the cycle progresses as the black hole shrinks and heats up. The black hole is evaporating. Eventually, the accelerating cycle gets out of control; the black hole shrinks down to nothing in the blink of an eye and disappears in a flash of radiation. The black hole dies.

It takes a very long time for black holes to evaporate. A small black hole twice the mass of the sun would take more than 10^{67} years to radiate itself away and explode; the universe, by contrast, is only a bit more than 10^{10} years old. But someday, many, many years from now, black holes across the universe may start exploding, one by one, as their event horizons shrink into nothingness. Releasing the information that had once been hidden. Maybe.

It seems likely that the evaporation and explosion of a black hole releases the information that was hidden behind the event horizon, kept from Nature's prying measurements. If the information is stored, rather than destroyed, it will be freed when the black hole dies, and the law of conservation of information will be held absolute; information will survive even a trip into a black hole. However, it is quite possible that the information is lost forever. If you dump a qubit into a black hole and the explosion doesn't release that qubit into the environment in some fashion, that qubit has been destroyed. Black holes would trump the law of information conservation. Nobody knows which scenario is true, and on February 6, 1997, three famous physicists made a wager about this very point. The terms of the bet were as follows:

Whereas Stephen Hawking and Kip Thorne firmly believe that information swallowed by a black hole is forever hidden from the outside universe, and can never be revealed even as the black hole evaporates and completely disappears,

And whereas John Preskill firmly believes that a mechanism for the information to be released by the evaporating black hole must and will be found in the correct theory of quantum gravity,

Therefore Preskill offers, and Hawking/Thorne accept, a wager that:

When an initial pure quantum state undergoes gravitational collapse to form a black hole, the final state at the end of black hole evaporation will always be a pure quantum state.

The loser(s) will reward the winner(s) with an encyclopedia of the winner's choice, from which information can be recovered at will.[7]

Hawking and Thorne wagered that black holes truly consume information, destroying it when it passes the event horizon. If you store a qubit on a star, say, a pure **(0)** or **(1)** or a mixed **(0 & 1)**, and that star suddenly collapses into a black hole, that qubit is lost to the universe forever. Preskill, on the other hand, gambled that information is conserved. Though the qubit is lost to Nature while the black hole exists, it is just trapped until the black hole explodes. When the black hole self-destructs and the event horizon disappears, the original qubit will be there, somewhere. If the star started out in a pure state, a **(0)** or a **(1)**, that pure state will, once more, be measurable. If the star started out in a mixed state, such as a **(0 & 1)**, the mixed state, too, will once more be measurable. The qubit was simply in deep storage; it wasn't destroyed. The law of information conservation holds. Though the Preskill-Thorne-Hawking wager seemed like a pointless bet—and one that might never have been resolved—they were wagering on nothing less than the fundamental laws that govern the universe. If information is not conserved, if it is destroyed by a black hole, scientists must look elsewhere for laws that hold everywhere in the universe. But if information can survive a trip into a black hole—indeed, it may be the *only* thing that seems to remain unchanged after crossing an event horizon—information might be the fundamental, immutable language of Nature. Physical laws, even those that apply at the center of the black hole, would obey the law of information. Information would be the supreme law.

But which side was correct? Information and Preskill, or a black hole and Hawking and Thorne? If you had wanted to lay down a side bet,

7. Preskill, "Black Hole Information Bet."

well, you missed your chance. In a much-ballyhooed announcement at a general relativity conference in Dublin in 2004, Hawking conceded the wager. He came up with a mathematical theory that, he argued, shows that information can't be irreversibly consumed by black holes. "If you jump into a black hole, your mass-energy will be returned to our universe ... in a mangled form which contains the information about what you were like, but in a state where it cannot be easily recognized," said Hawking, who then handed Preskill a copy of *Total Baseball: The Ultimate Baseball Encyclopedia*.[8] (Thorne did not concede the bet, as he is not yet convinced; he agrees to pay Hawking back if and when he finally comes around. Ironically, Hawking's mathematical theory seemed not to have changed any minds but Hawking's.)

When Hawking conceded the bet, the most vocal opponent of information conservation had thrown in the towel. It took decades of argument about black hole thermodynamics, general relativity, particle physics, and information theory to convince Hawking—and there are still some holdouts, even though most of the particle physics and string theory community had long since become convinced that information could survive even the ultimate destructive power of a black hole.

One of the most compelling reasons why physicists believe that information is always conserved has to do with the fact that the thermodynamics of black holes implies that they have not only temperature but also entropy. The Boltzmannian laws that describe the arrangements of atoms in a gas—and led to information theory—also apply to a black hole.

When matter falls into a black hole, it loses its identity. Throw one kilogram of hydrogen or one kilogram of feathers or one kilogram of lead or one kilogram of antimatter or one kilogram of kittens into a black hole and the end result will be the same. The black hole gulps the

8. Author's notes of Stephen Hawking's presentation, 21 July 2004.

matter down and expands a little. Its event horizon increases in area a small amount, and the information about the matter that was dumped in the hole is lost to Nature.

There are a huge number of things we can dump in the black hole to get the same result. There are an enormous number of ways we can get the black hole to increase its area in that particular manner, yet the black hole that has swallowed a kilogram's worth of lead is indistinguishable from one that has gulped down a kilogram's worth of feathers. In other words, there is a *degeneracy* between a black hole that has gobbled the feathers and one that has gobbled the lead. Way back in chapter 2, the discussion of entropy began with throwing marbles into a box—and since those marbles were identical, many of the arrangements were degenerate with one another. The indistinguishability of those arrangements led to the bell curve, which, in turn, led to the concept of entropy.

In the 1970s, scientists such as Hawking, Thorne, Wojciech Zurek, and Jacob Bekenstein realized that the process of dumping matter down the throat of a black hole is exactly analogous to dumping marbles in a box. Both cases lead to the concept of entropy. The mathematics is quite similar to the case of a container full of gas. It turns out that a black hole's entropy is proportional to the logarithm of the number of ways it could have been made. $S = k \log W$. A black hole is subject to the laws of thermodynamics just as a container full of gas is.

But there's an interesting wrinkle for a black hole. Dumping matter into a black hole increases the black hole's entropy. It also increases the event horizon's area by a certain amount. It turns out that these two properties—entropy and event horizon area—are inextricably linked. Increase one and you increase the other by the same amount; decrease one and you decrease the other in the same proportion. A black hole's entropy is exactly the same thing as the size of its event horizon.

If a black hole has entropy, perhaps it, like a container full of gas, has a number of different configurations it can be in. Though a black hole is externally featureless, the black hole itself might have an enormous number of different quantum states. It might be capable of storing qubits—and the number of qubits it would be storing would also be proportional to the surface area of its event horizon.

Scientists don't really know how to describe a black hole in quantum-mechanical terms, so they don't know the details yet about whether or not information can reside in a black hole. But there are a bunch of theoretical results that are encouraging. String theorists have ideas about how information can be preserved in a black hole. So do scientists who adhere to another type of theory: loop quantum gravity. Other techniques, such as treating the black hole like a giant vibrating atom, also give hints about the quantum nature of black holes. Recently, quantum loop gravity and the vibrating-atom technique gave a remarkably similar picture of space and time around a black hole, perhaps suggesting that scientists are on the right track to understanding black hole physics.

That track is leading many scientists to think that a black hole can store information. Indeed, most scientists believe nowadays that you can talk about the information content of black holes, and that the information inside a black hole is related to the size of its event horizon. Some go even further and argue that a black hole can *process* information. In 2000, the MIT physicist Seth Lloyd set out on a whimsical quest to design the ultimate laptop—the fastest possible computer. In his thought experiment, he tried to figure out the largest number of computations a one-kilogram mass—in any configuration—could make in a second. He calculated that if it were confined to the space of a liter, a one-kilogram mass could store and manipulate about 10^{31} bits of information. Then he figured out how quickly a laptop could manipulate those bits.

Heisenberg's uncertainty principle is the limiting factor; the energy-time relationship means that the faster you manipulate a bit of information, say, flipping a **0** into a **1** or vice versa, the more energy you need to flip that bit. So, to make the computer as fast as possible, Lloyd converted all of his ultimate laptop's mass into energy via Einstein's equation $E = mc^2$. The mass becomes a billion-degree ball of plasma with an enormous amount of energy available to process the information it contains. Of course, this would make it very difficult to package Lloyd's laptop, but no matter.

But speeding up bit flips and increasing processing speed is only half the story. If you *really* want to speed up your computer, you must also slash the time it takes for memory locations to communicate with one another. Since the information in the computer is physical and must be transported from place to place at the speed of light or slower, the less distance the information has to travel, the faster the computer can perform its operations. So Lloyd imagined compressing his plasma laptop down into the smallest space possible: he compressed it into a black hole. This minimizes the time it takes for information, which would presumably reside on the event horizon, to go from place to place. (None of the information that the black hole processes goes back outside the event horizon, making readout impossible, but that doesn't stop the black hole computer from doing its job, because you can freely send information from point to point on the surface of the event horizon.)

When Lloyd did the calculations, he was surprised to find out that the time it took for parts of a one-kilogram black hole to send information to other parts of the black hole was *precisely* the same time it takes to flip a bit with one kilogram's worth of mass-energy. No time is wasted in bit flipping and none is wasted in communication; the two processes take exactly the same amount of time. Perhaps it's not a coincidence that these two different things have the same value. Perhaps a black hole really *is* the ultimate computer, the ultimate processor of information. If so, it would be a ringing confirmation that information is the way to plumb the depths of the black hole. Information is supreme. It might even reveal the existence of hidden universes.

CHAPTER 9

COSMOS

Behold! Human beings living in an underground den, which has a mouth open towards the light and reaching all along the den; here they have been from their childhood, and have their legs and necks chained so that they cannot move, and can only see before them, being prevented by the chains from turning round their heads. Above and behind them a fire is blazing at a distance, and between the fire and the prisoners there is a raised way; and you will see, if you look, a low wall built along the way, like the screen which marionette players have in front of them, over which they show the puppets....

You have shown me a strange image, and they are strange prisoners.

Like ourselves.... To them, the truth would be literally nothing but the shadows.

—Plato, *The Republic*

The universe runs on information. At the smallest scales, Nature is constantly making measurements, collecting information, and disseminating that information into the environment. As stars are born, shine, and die, their information is scattered throughout the galaxy; as black holes devour all matter and energy that strays too close, they are devouring information—perhaps, in a sense, becoming the ultimate computer.

But our picture of the universe is not yet complete, not by a long shot. Scientists don't understand the structure of the universe on a philosophical—or a physical—level. They don't know whether ours is the only universe or whether there are others out there inaccessible to us. They don't understand the mechanisms that make quantum mechanics so weird; they don't truly know *how* entangled particles are able to conspire with each other despite the lack of information exchange between them. They don't understand the structure of space at the tiniest scales, and they don't understand the nature of the universe at its largest scales.

Information theory does not provide the answers to these questions just yet, but it is yielding clues to all of them. Not only is information providing a glimpse of a region of space totally inaccessible to experiment—the inside of a black hole—it is revealing the structure of space and time. In the process, it implies the presence of entire universes parallel to our own, unseen and unseeable. While these parallel universes stretch the credulity of even their advocates, they can explain the big paradoxes in quantum mechanics. Parallel universes reveal how superposition works, and how distant entangled particles can instantly "communicate" with each other over vast distances. The mysteries of quantum mechanics become much less mysterious—once you believe that information creates the structure of space and time.

It's an unsettling idea. The frontiers of information theory are providing a very, very disquieting picture of our universe—and of the ultimate fate of life in the cosmos.

Black holes are, in some ways, universes unto themselves. Remember that probe we sent into the black hole in chapter 8? What if it found life? If there were some sort of creature that was able to make its home inside an event horizon, it would be able to see all the stars and galaxies in the sky above. It might even be aware of the small, bluish planet that we live on. However, no matter how hard it tried, the creature would be utterly unable to send us a message. Whatever information it tried to send, whatever message it attempted to beam to us, would never cross the event horizon. The pull of the black hole is too strong. Even if there were a huge population of these creatures swirling around the black hole, all screaming and signaling as loud as they possibly could, Earth would never receive a single bit or qubit of information about them. Their information is simply inaccessible to us. For all we know, a whole universe of stuff is lurking behind the event horizon of a black hole—a universe that we're unaware of because we are unable to gather information about it.

Of course, this is a purely hypothetical speculation. It's unlikely that there are black hole creatures or other universes on the other side of event horizons. Yet the event horizon shows that it is possible that there can be real things out there that aren't truly part of our universe. There could even be stars and galaxies and creatures that are cut off from us by some sort of barrier that blocks information; there could be objects in our cosmos that lead an existence entirely separate from ours. It would be impossible, even in theory, to set up a dialogue between ourselves and the creatures in such a place. In some sense, if you set up an information blocker between two regions of space so that they can't communicate with each other, the two become, essentially, different universes.

It's an odd idea. After all, *universe,* by definition, includes everything in ... well, in the universe. But scientists have begun to consider the idea that there are alternate, separate universes from our own. In fact, a good number of physicists take the idea seriously. Some even believe that alternate universes *must* exist; they may be an inescapable consequence of the laws of information and of the physics of black holes.

The first step in the road to alternate universes has to do with what happens to information in black holes. In the last chapter, we saw that the information that a black hole swallows seems to be related to the surface area of the black hole's event horizon. As a black hole gobbles more and more matter and energy—more and more information— the surface area of the event horizon grows. Indeed, the black hole's entropy is proportional to the area of its event horizon—the surface area of the event horizon divided by 4, to be precise. It doesn't matter if the black hole is perfectly spherical (and so encloses the maximum amount of volume possible) or somewhat flattened by the its spin (and so encloses a bit less volume): the information content of black holes—if, indeed, information is preserved—is the same if the surface areas of their event horizons are the same.

This is a fairly uncontroversial belief now. Most scientists accept that you can talk about black hole information, and that this information is proportional to the area of its event horizon. But this belief has a very, very strange consequence when a black hole swallows information. The strangeness has to do with the difference between an object's volume and its area. When you heft a heavy object, such as a brick of lead, you are making a rough measurement of the amount of matter that is in the brick. The heavier the object is, the more mass the brick has—how much "stuff" is in the brick. The mass of the brick, in turn, is related to its volume. You could increase the chunk of lead's surface area—you could pound it flat—or you could decrease the chunk's surface area— you could shape it into a ball—but the mass remains the same because the *volume* of the chunk doesn't change. It is volume, not area, that is the gauge of the amount of stuff in an object. If you are storing information (or quantum information) in that chunk of matter, you would expect it to be proportional to the amount of stuff in that chunk: you'd expect it to be proportional to the matter's volume, not its area.

But with a black hole, the situation is exactly the opposite of what you would expect. It's as if the information in a black hole "lives" on the event horizon's surface area rather than in the volume that the event horizon encloses. The amount of stuff in a black hole is proportional to its area, not its volume. This is quite odd. The surface of an event horizon is really a two-dimensional surface, like the skin of a hollow, infinitely thin-walled ball. It's not truly a three-dimensional object like a solid sphere. This means that all the information in the black hole resides in two dimensions rather than three.[1] It is as if

1. Einstein's formulation of the theory of relativity treats time as another dimension. Our universe is therefore four-dimensional, and the event horizon is three-dimensional. For simplicity's sake, I will stick to the more familiar two- and three-dimensional objects, especially since some theories, like string theory, take us up to ten or eleven dimensions.

information completely ignores one of our dimensions. In a sense, the information is like a *hologram*.

A hologram is a peculiar sort of image you're almost certainly familiar with already. Most Visa and MasterCard credit cards sport one as a security feature. It's that peculiar floaty-looking picture rendered in foil on the front of the card. It's not so easy to see with cheap, low-quality holograms like the ones on credit cards, but if you look carefully at the image, you might notice that it appears three-dimensional. It seems as though it is floating in space.

Holograms exploit the wavelike properties of light to make a special kind of photograph—a three-dimensional picture—of an object. Even though the hologram is stored on a two-dimensional substrate such as a flat piece of film or foil, the hologram encodes the full three-dimensional information about the object that was imaged. With a high-quality hologram, such as the ones you see in many science museums, a flat piece of film really does produce a truly three-dimensional image of a pair of dice or a skull or some other object. Walk around the hologram and you will see different faces of the dice or different bones in the skull, something that would be impossible with an ordinary two-dimensional photo. In a hologram, all the three-dimensional information about an object can be stored on a two-dimensional piece of film.

A black hole, like a hologram, seems to record three full dimensions' worth of information—all of the (three-dimensional) stuff that fell past the event horizon—upon a two-dimensional medium— the surface area of the black hole's event horizon. In 1993, the Dutch physicist Gerardus 't Hooft (who won the Nobel Prize in 1999 for different work) proposed what is now known as the *holographic principle*, which, for fairly solid theoretical reasons, extends the physics of black holes to the entire universe. If the principle is correct, then, in a sense, we might be holographic ourselves: two-dimensional creatures that are merely laboring under the illusion that we are three-dimensional.[2] It's

2. Or, more precisely, three-dimensional creatures that labor under the illusion that we are four-dimensional. As if this idea weren't strange enough.

a very odd possibility, but nobody knows whether the holographic principle is correct. Even if it isn't, though, information theory has another surprise in store.

On firmer ground than the holographic principle is that a finite-size chunk of matter can store a finite amount of information. A black hole, which, after all, is the densest possible chunk of matter (and, in the abstract, the ideal information-processing machine, as Seth Lloyd showed), has information content proportional to the surface area of its event horizon. So long as the black hole's mass is finite, its event horizon is finite. If its event horizon is finite, then the amount of information it can contain is finite—and proportional to the surface area of the event horizon that surrounds it.

In 1995, the physicist Leonard Susskind proved that this is true not only for black holes but for all matter and energy, no matter what its form. If you can take a hunk of matter and energy and surround it with an imaginary sphere with surface area A, the amount of information that matter and energy can store is at most $A/4$, in the appropriate units. This is known as the *holographic bound,* and it is a consequence of the laws of information and thermodynamics.

According to the holographic bound, even a small chunk of matter can theoretically store an astronomical amount of information. (A dollop of matter a centimeter across can, in theory, store up to 10^{66} bits—a mind-bogglingly huge number roughly equivalent to the number of atoms in a galaxy.) However, that number is *finite,* not infinite. If you can contain a section of the universe with a ball of a finite surface area, it can only contain a finite amount of information, even when the ball is absolutely enormous. This is on very solid theoretical ground—you have to accept it if you accept that the second law of thermodynamics holds for black holes—but it leads to some bizarre conclusions.

Information is physical. It is not an abstraction that miraculously sits on an atom or an electron; the information must be stored on that object and the information must manifest itself in some physical manner. You can store a qubit on an atom by manipulating that atom's spin, or its position, or some other physical attribute of the atom, and each qubit you store must be reflected in the overall properties—the quantum states—of the atom. This is not news to you; from the example of Schrödinger's cat on, we've been exploring the relationship between an object's quantum state and the information that the quantum state represents. But scientists argue that if there's only a finite amount of information in a given chunk of matter, any object made of matter must be in one of a finite number of possible quantum states—in other words, an object can only have one of a finite number of quantum wave functions, where the wave function encodes all of an object's information, accessible or inaccessible. Thus, if you imagine a ball with a finite surface area, theorists contend that there are only a finite number of ways that matter and energy can be arranged inside it.

This is easier to see if we look back at our previous analysis of black holes. The surface area of a black hole's event horizon represented the information that the black hole swallowed. What, precisely, did that information represent? Well, once you dump matter down a black hole, you lose all information about what kind of matter it is; you don't know whether it was atoms or neutrons or Ford Pintos, much less whether the Ford Pintos were painted red or blue, or whether the atoms were spin up or spin down or both at the same time. In other words, you lose all the information about the nature of the matter you dump in the black hole; you lose all the information about the matter's quantum states. But the information that you lose is stored by the black hole (if information is truly conserved) and goes into increasing the area of the event horizon. So, the information on the event horizon is equivalent to the information about the quantum states of the matter that you dump in the black hole. Information—quantum states—area. All three are linked.

So far, so good. Within a finite ball, there are only a finite number of ways to arrange the stuff inside. But things start to get downright silly when you start considering really large balls—as large as the visible universe. The universe is only about 13.7 billion years old, and light began streaming freely throughout it a little less than 400,000 years after the big bang. That light is the most ancient light we can see. It is the edge of the visible universe; anything beyond that is invisible. Since information travels no faster than the speed of light, if you draw an invisible, enormous (but finite) sphere around the Earth that has a radius of tens of billions of light-years, you encompass all of the universe that has been able to send information to us since the moment when light was set free.[3] Conversely, everything that can possibly have received information about the Earth since that time is contained within that sphere. In other words, every element of the universe that can swap information with us since that era 400,000 years after the big bang is encompassed in an enormous but finite sphere. For the sake of brevity, let's call this sphere our Hubble bubble.

There's probably more to the universe than our Hubble bubble. Scientists are almost certain that there is more to the universe than what is visible—more than what is encompassed in that giant sphere. Indeed, most cosmologists think that the universe is infinitely large. At the moment, scientists believe that our universe is infinite in extent— that it has no borders—and that it doesn't have a funky shape that curls around on itself, as a handful of scientists have unconvincingly argued. If you take a rocket ship and travel in one direction for years and years and years, you will never come across an uncrossable boundary and you will never revisit the place you set off from.

---

3. Don't worry if this doesn't seem to make any sense, but the radius of that sphere is actually somewhat bigger than 13.7 billion light-years. This is because the fabric of space is constantly expanding. If, 14 billion years ago, we had a snapshot of the universe, we could draw a 14-billion-light-year circle around the point in space that will eventually become the Earth, and anything within that sphere will be causally connected with the Earth 14 billion years later. But the fabric of space and time is not a snapshot; 14 billion years later the sphere has expanded to a radius of about 40 billion light-years. We are receiving light from objects in that 40-billion-light-year sphere, even though the universe is less than 14 billion years old. (It's a weird consequence of relativistic mathematics; remember, it comes at us at 300,000,000 kilometers per second no matter how the Earth is moving—and that includes the motion due to the expansion of spacetime.) However, it's not terribly relevant whether the sphere is 14 or 40 or 6 zillion billion light-years in radius. All that matters is that the sphere is finite.

Physicists don't use the term *infinite* lightly, but they came to the conclusion of an infinite universe for a number of reasons. For one thing, astronomers have been trying to see hallmarks of a finite universe and have failed. For example, when cosmologists looked at the ancient cosmic radiation from 400,000 years after the big bang, they saw that the lack of patterns in that radiation imply that our universe has a radius no smaller than 40 billion light-years—there's no sign, yet, of any edge to the universe. Though this is one piece of evidence for an infinite universe, it is not what really makes physicists think that the universe is infinite. The real motivation for a never-ending universe is the theory of inflation.

Inflation is a very successful cosmological theory that describes the universe in the first few fractions of a second after the big bang, and it seems to imply that the universe is infinite in extent.[4] Of course, inflation could be wrong on some level (even though it seems to be working). Alternatively, inflation could be entirely correct, yet the interpretation that it leads to an infinite universe might still be mistaken (even though the mathematics seems to point in that direction). But at the moment, most cosmologists consider the universe to be infinitely large. Combined with the holographic bound, this means big trouble.

If the universe is infinite, our Hubble bubble, which is finite in extent, is just one of many, many, many nonoverlapping Hubble-bubble-size spheres you could draw in the universe: the universe can have a huge number of independent Hubble bubbles. Indeed, since our Hubble bubble is finite, in an infinite universe, you could fit an *infinite* number of these independent Hubble bubbles in the universe. Now the information-theoretic catch: each of those spheres has a finite surface area, so each has a finite information content, a finite number of quantum states, and a finite number of ways that matter and energy

---

4. Though the details of inflationary theory are beyond the scope of this book, interested readers might consult my book about cosmology, *Alpha & Omega*.

can be arranged within each Hubble bubble. There are only a finite number of wave functions that the stuff inside each Hubble bubble can have.

The wave function captures every single piece of information about all the stuff—all the matter and energy—in our Hubble bubble, whether we are aware of it or not. It encodes the location and momentum of every single atom in that Hubble bubble, as well as everything else you can possibly imagine about our bubble. In it are encoded the position and color of every lightbulb on Piccadilly Circus, the velocity of every fish in the sea, and the contents of every single book that exists on Earth. Our Hubble bubble's wave function even includes *your* wave function; it encodes every single morsel of information about you, down to the quantum states of each atom in your body. Though this is an unbelievably large amount of information, our Hubble bubble's wave function contains everything about our visible universe. Just for the heck of it, let's call it wave function #153.

There are only a finite number of wave functions for a Hubble volume. There are an unbelievably, unbelievably huge number of possible wave functions (call it a *kergillion*), but that number is finite nonetheless. So our wave function is one of a kergillion possible wave functions. Other than the fact that it is *our* wave function, there is probably nothing particularly special about it. It's probably not all that much more probable or improbable than the other kergillion possible wave functions.[5]

But remember, there are an *infinite* number of these Hubble bubbles in an infinite universe. Infinity is more than a kergillion—even more than a kergillion plus one. And once we reach a kergillion-plus-one Hubble bubbles, something incredible must have happened. There are only a kergillion possible wave functions a Hubble bubble can have, so in a collection of a kergillion-plus-one Hubble bubbles,

---

5. Actually, it doesn't really matter how improbable our particular wave function is; the following argument holds so long as wave function #153 is not *impossible*.

there must be *at least* one duplicate! Two Hubble bubbles must have *exactly* the same wave function. Every atom, every particle, every little dollop of energy is in exactly the same place, has exactly the same momentum, and is exactly the same in every single possible way you can imagine—and even in the ways that you can't imagine.

Why stop at a kergillion plus one? At a kergillion-plus-two Hubble bubbles, there must be two duplicates. At two kergillion Hubble bubbles, there are a kergillion duplicates: on average, there are two copies of every possible wave function. At a million kergillion Hubble bubbles, there are, on average, a million copies of every single possible wave function. Including wave function #153. Ours.

If there's nothing particularly special about our wave function, then in a volume that contains a million kergillion Hubble bubbles, there are about a million *identical* copies of our universe. There are a million copies of Hubble bubbles that are identical down to the position and color of every lightbulb on Piccadilly Circus, the velocity of every fish in the sea, and the contents of every single book that exists on Earth. Each of those Hubble bubbles even contains an identical copy of *your* wave function—down to the quantum states of each atom in your body. There are a million copies of you, identical in every detail. In fact, those million doppelgangers are reading a doppelganger copy of this book and are finishing this paragraph as you are, right . . . now.

Indeed, if the universe is infinite, then physicists estimate that an identical Hubble bubble should be roughly $10^{10^{115}}$ meters away from us. Of course, you'd never be able to communicate with your doppelganger, as it would be vastly, vastly more distant than the edge of our visible universe, but if the universe is infinite, that doppelganger should be there nonetheless.

But wait! It gets weirder! The finite number of wave functions were caused by the finite information that could be stored within a given volume, and this in turn implied a finite number of possible configurations of mass and energy. Each possible configuration of mass and energy and information was assigned a wave function; each was given a number. And each was counted among the kergillion possibilities. Therefore, our collection of a kergillion wave functions contained every possible configuration of matter and energy that a Hubble bubble could have. And in our collection of a million kergillion Hubble bubbles, on average, there are a million of each.

Is it possible to have a universe that is populated by a race of superintelligent octopuses? Got a million of those. Is it possible to have a universe where everyone on Earth communicates via an intricate language of tap dancing and flatulence? Got a million of those. Is it possible to have a universe identical to our own except for the fact that you are reading this book in pig Latin? Got a million of those. If the universe is infinite, then every single conceivable configuration of matter in a Hubble bubble that is not forbidden by the laws of physics *must* exist somewhere. In a sense, our cosmos would be composed of many independent parallel universes, each of which can take one of a finite number of configurations.

Of all the insane things that I have tried to convince you of in this book, this is by far the craziest. I, myself, have a very, very hard time believing it. I'd like to think that there's a flawed assumption somewhere—something that physicists have gotten wrong or have overlooked. But the logic seems fairly airtight. If the universe is infinite, and if the holographic bound is correct, then it is hard to escape the idea that the cosmos is populated with infinite copies of you—and, worse still, there are also infinite copies of you getting eaten by a giant carnivorous alien wombat (and vice versa).

If you go up to a physicist in the field and ask him about this, he'll probably hem and haw and avoid the question. But quite a number of eminent, noncrazy physicists will say, quite confidently, that they believe that identical or nearly identical copies of themselves are floating out there in the cosmos—even if they don't necessarily buy the argument I've set forth above. There's a very different reason physicists have for believing that parallel universes exist. This, too, has to do with information theory and the laws of quantum theory. Scientists are beginning to accept a slightly different flavor of parallel universe— one where information quite literally shapes the cosmos—and in the process they are resolving the problems of quantum theory.

In 1999, roughly one hundred physicists took an informal poll at a quantum computation conference. Thirty of them said they believed in parallel universes or something quite similar, even though no direct evidence has been found that such universes exist. This belief is, in large part, a consequence of the mysteries of quantum theory. Information theory is yielding a great deal of insight into those mysteries, such as how quantum objects behave; by studying the exchange of information among objects, observers, and the environment, physicists are learning about the laws of the quantum world. But information is not enough. Something is still missing. Quantum theory is not complete.

The mathematics of quantum theory is incredibly powerful. It makes predictions with incredible precision, and it does a superb job of explaining how particles behave. However, that mathematical framework comes with a great deal of philosophical baggage. The math of quantum theory tells how to describe an object in terms of its wave function, but it doesn't tell you *what* that wave function is: Is it a real object, or is it a mathematical fiction? The math of quantum theory describes objects' behaviors with the phenomenon of super-position, but it doesn't explain *how* superposition works or how it collapses: What does it mean for an object to be in two places at once, and how can that property suddenly disappear? The math of quantum theory explains the spooky action at a distance between two entangled particles, but it doesn't explain *how* two distant particles can conspire with each other without passing information back and forth. The math of quantum theory is very clear. The physical reality that quantum theory is describing is *far* from clear.

A scientist can get away with ignoring reality. If the mathematics works and it predicts the physical phenomena you are studying, you can just pay attention to what the equations are saying without trying to figure out what those equations mean. (In a phrase attributed to the Nobel laureate Richard Feynman, this can be called the "shut up and calculate!" attitude.) But most physicists are convinced that the numbers they deal with are reflections of a genuine physical reality out there. And most of them want to know what physical reality their mathematics represents. It is not enough to have a mathematical description of a phenomenon; they want to know about the physical processes that their equations describe. They want to know how to *interpret* their mathematical framework. This is where the trouble really resides.

Though mainstream scientists tend to agree about all the mathematical conclusions of quantum mechanics, they disagree about the interpretation of what those conclusions actually mean in reality. There are a number of schools of thought—a number of interpretations of how the mathematics of quantum theory reflects physical reality. How is it that quantum mechanics—and experiment—says that an object can be in two places at once, yet as soon as we attempt to observe the superposition, it is destroyed? What, physically, is going on?

One interpretation, one way of explaining how particles can exist in superposition and how entangled particles can communicate, relies upon information and parallel universes to explain the weirdness of quantum theory. However, this is not (yet) the standard interpretation of quantum mechanics. That honor goes to what is known as the Copenhagen interpretation. Created by some of the founders of quantum theory in the 1920s, including Copenhagen resident Niels Bohr and the German Werner Heisenberg, he of the uncertainty principle, the Copenhagen interpretation answers the question by giving a special role to observations. The wave function of, say, an electron is really a measure of the probabilities that an electron will be found in a certain place. So long as the electron remains unobserved, this wave function evolves smoothly. Like a fluid, it can spread out, flowing into several different regions at the same time; it can evolve into a state of superposition. But as soon as an observer makes a measurement and tries to find out where the electron is, collapse! The wave function somehow instantly ruptures, quivers, and shrinks. A celestial coin flip determines where the electron actually *is* in space; the electron "chooses" its position according to the probability distributions that the wave function described.

For many years, the Copenhagen interpretation was the only game in town, but it had some troubling aspects to it. For one thing, the act of observation was ill defined, a problem that was largely responsible for the trouble with Schrödinger's cat. The Copenhagen interpretation didn't really address the meaning of *observation*. Observations tended to be phrased in terms of a conscious being's making a measurement, but need the observer truly be conscious? Would a scientific instrument cause a wave function to collapse, or does the instrument need to be seen by a conscious scientist before the collapse occurs? The Copenhagen interpretation left this open, as it did the question of how, precisely, the collapse process occurs. It answered neither when nor how the collapse of a wave function happens. Nor did it really answer whether the wave function is, on some level, a physical object or whether it is a mathematical fiction that has no true physical analogue. Though the wave function says so, is the electron *really* in two places or not? Copenhagen doesn't tell you. Because of the huge questions left unanswered in the Copenhagen interpretation, you can have two physicists who both claim to believe Copenhagen, yet have very different views on the nature of reality. One might think that the wave function is real and electrons truly can be in two places at once, while the other doesn't believe either. This is a highly unsatisfying situation, to say the least.

In the 1950s, a number of physicists proposed other interpretations to address the problems with Copenhagen. Because of this, there are several other interpretations of quantum mechanics that have a very different point of view. All of them create as many problems as they solve; they usually propose some radical phenomenon that is just as ridiculous as the bizarre behavior that they are trying to explain away. But if you're to get beyond "shut up and calculate!" and have some sort of understanding of the physical reality, you've got to use one of these interpretations to try to make sense of what the math is saying. This book is no exception—it is not exempt from the drawbacks of these interpretations.[6]

However, one interpretation is rapidly becoming a favorite of physicists. Like the other alternatives to the Copenhagen interpretation, it carries a large burden with it—a radical and counterintuitive phenomenon. But it is no more radical than the conclusion of the argument I've set forth above: there are parallel universes. If you accept that possibility, then quantum theory begins to make physical sense, and information becomes a fundamental part of the fabric of space and time. This solution was born in 1957, when a graduate student at Princeton, Hugh Everett, proposed an alternative to Copenhagen that became known as the *many worlds interpretation*. The core of Everett's argument is that the wave function is a real object, and when it says that an electron is in two places at once, it really is in two places at once. But unlike all variants of the Copenhagen interpreta-

6. Throughout this book, I have been using vocabulary from whatever interpretation makes it easiest for me to communicate the point I'm making. The result is something of a hybrid, a cross between a Copenhagen interpretation where the wave function is considered a real object and a many worlds interpretation. Even though you might choose a different interpretation from the one I've been using, it is largely irrelevant to the phenomena I talk about in the book. There's no way to distinguish which interpretation is "right"; they're nearly equivalent in their predictions, and *completely* identical when it comes to experiments that have been performed in the past and are likely to be performed in the near future. You might disagree with my bold statement that an electron can be in two places at once—you might believe that there's only a single electron and it is a "pilot wave" that is in two places at once—but the outcome of all the experiments I describe will be precisely the same. Furthermore, *all* the interpretations agree that there's a fundamental difference between the classical world and the quantum world; they all show how it's impossible to explain, say, the two-slit experiment with a single, classical object going through a single slit without creating some radical new mechanism to explain how it can interfere with itself.

tion, there is no real "collapse" of the wave function. When information leaks out about an electron in superposition, when someone measures whether the electron is on the left or on the right, the electron chooses . . . both. It gets away with this in a very odd manner—by altering the structure of the universe with the assistance of information.

To envision what is going on in the many worlds scenario, it helps to think of our universe as a thin, transparent sheet like a strip of celluloid. An object in superposition sits happily on that sheet, existing simultaneously in two places at once, perhaps creating an interference pattern. When an observer comes along and gathers information about the particle, say, by bouncing a photon off it, the observer will see the electron in either the right position or the left position, not in both at the same time. A Copenhagenist would say that the wave function collapses at that point; the electron "chooses" to be on the right or on the left. A many worlds adherent, on the other hand, would say that the universe "splits."

A godlike being, watching the interaction from outside the universe, would suddenly realize that the celluloid universe that the electron (and observer) inhabits is not a single sheet, but two sheets stuck together. When information leaks out about the electron's position, it is really yielding information about the structure of the universe: the information shows that the universe is twofold. In one of these universes, the electron inhabits the right position; in the other, the electron inhabits the left position. As long as these sheets are stuck together, it is as if the electrons are on the same sheet; the electron is in two places at once and interferes with itself. But the act of gathering information about the electron's position peels the sheets away from each other and reveals the true multifoliate nature of the cosmos; the sheets diverge because of the transmission of information.

While a godlike being would be able to see these universe sheets divide, the observer who made the measurement, also embedded in those sheets, would be totally unaware of what was happening. And trapped on the sheet, this observer would also be split in two, Observer Left and Observer Right. Observer Left, on his sheet, would see the particle on the left; Observer Right, on his sheet, would see the particle on the right. And since the two sheets are no longer in contact with each other, the two copies of the particle and the two copies of the Observer are no longer able to interact with each other. They now inhabit separate universes. Though the godlike being would be able to see the full, complex, multileaved structure of these parallel universes—the *multiverse*—an observer in that universe would still think

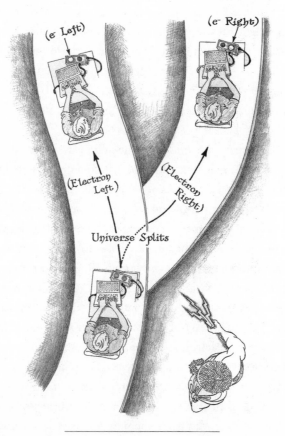

Superposition in the multiverse

he inhabits a single sheet, completely unaware of the alternate universe where the measurement had the opposite outcome. And once the two leaves split from each other, they are essentially unable to communicate; they cannot exchange information. It is as if there is a barrier between the two leaves. In essence, the two are in different universes, even though they are in the same multiverse.

This idea—a multiverse that splits owing to the exchange of information—also provides a nice explanation for the spooky action at a distance. Take an EPR pair of particles, say, entangled in terms of position. If one is on the left, the other must be on the right, and vice versa. But if you create them in a state of superposition, neither particle "chooses" whether it's on the left or on the right until it is measured; each is an indeterminate mix of left and right until the act of measurement—until something (Nature or an observer) gathers information about each particle.

Take an EPR pair entangled in this way and send one toward an observer on Earth and one toward an observer on Jupiter. Each observer makes an observation when the particle arrives; each gathers information about the state of the particle, splitting the *world-sheet*—and each observer—in two. But the splits are *local* splits. A godlike being would see the world-sheet divide near each of the two observers, but in between the two observers the sheets would remain stuck together. Only when one of the observers (say, the Earth observer) sends a bit of information to the other (the Jupiter observer) do the sheets in between the two begin to peel apart. The dollop of information, which moves at most at light speed, splits the universe as it travels. When it reaches Jupiter, it completes the separation; the two world-sheets are completely separate. In one of these now-separate sheets, the Earth observer has measured left and the Jupiter observer has measured right; in the other, the reverse has happened. In both cases, it seems as if the particles conspired with each other; even though no information travels faster than the speed of light, the two particles are always in opposite position: one is left and the other is right.

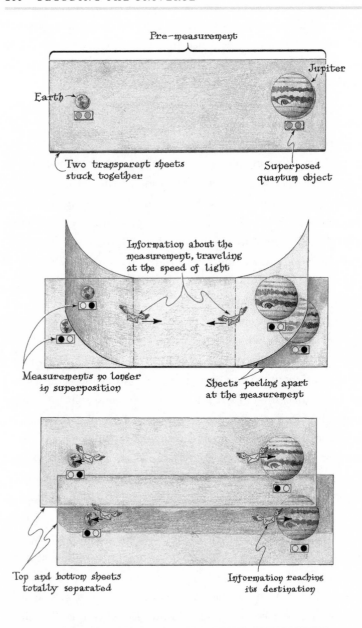

An EPR pair in the multiverse: information,
zooming at the speed of light, tears the sheets apart

To an observer embedded in the sheet, it appears that the particle only "chose" its position at the very moment of measurement; if one of the observers wanted, before a measurement, he could have seen an interference pattern that proved that the particle was in two places at once. A godlike being would see that the transfers of information just revealed the structure of the multiverse, peeling world-sheets away from each other and exposing its multifoliate nature. To someone embedded in the universe, such as a scientist in a laboratory, he would have to explain the bizarre phenomenon of two particles that don't "choose" their positions until the very moment of measurement, yet manage to conspire, across great distances, to be in opposite positions. Even the spooky action of entanglement makes physical sense in the many worlds interpretation.

It's a rather pleasing explanation. It's relatively simple, fairly neat, and all (all!) it requires is a belief in a multileaved multiverse instead of a single-sheeted universe. In the many worlds picture, a godlike observer would see the multiverse in its full complexity—a sheaf of world-sheets, stuck together in some places, separate in others. As information moves back and forth in the universe, it causes sheets to separate from each other, making the multiverse bubble and branch out. (With a reversible measurement, one where no information is dissipated, the sheets can even fuse back together.) Each measurement, each transfer of information—including those done by Nature—causes the multiverse to spread its sheets apart and flower. Information is what determines where the multiverse branches and where it fuses, where it spreads apart and where it is stuck together. In the words of the quantum physicist David Deutsch, "The structure of the multiverse is determined by information flow." Information is the force that shapes our cosmos.

But how radical is it to assume that there are parallel universes? Even with all the complexity of the multiverse, the incredible number of world-sheets is no more complex than the multitude of parallel universes that was postulated in the previous section. The same arguments apply. There are a finite number of possible wave functions that any given region of space can have; there are a finite set of possibilities for the configuration of energy and matter and information in a finite volume. Each sheet in the multilayered multiverse represents one possible configuration of matter and energy and information, and in an infinite universe all these configurations occur and recur and recur countless times at different regions in the universe.

Though the multiverse is extremely complex, the multiverse's parallel universes are no more complex than the ones that scientists think must exist in an infinite universe. So the radical phenomenon of many worlds isn't that radical after all. If you accept the conclusion that parallel universes exist—as many scientists have—you get, at no added cost, an explanation for all the bizarre phenomena in quantum mechanics. Superposition and entanglement no longer require a deus ex machina of a wave function "collapsing" or a particle "choosing." It is all a function of the information flowing from place to place, altering the structure of the multiverse in the process. Underneath it all, our universe may be entirely shaped by information.

Life, too, is shaped by information. All living creatures are information-processing machines at some level; intelligent, conscious creatures are processing that information in their minds as well as in their cells. But the laws of information put limits on the processing of information. There are a finite (if enormous) number of ways information can be arranged in our Hubble bubble, so there are a finite (and smaller, but still enormous) number of ways information can be arranged and processed in our heads. While humans might be able to contemplate infinity, we can only do so in a finite number of ways. The universe might be infinite, but we are not.

Indeed, all life in the universe must be finite. As the universe expands and evolves, the entropy of the cosmos increases. Stars burn out and die, and energy gets harder and harder to find. Galaxies cool down, getting ever closer to frigid equilibrium. And in a universe approaching equilibrium, it's hard to find energy and shed entropy; it gets more and more difficult to preserve and to duplicate your information. It becomes more and more difficult to sustain life. Must life die out completely?

In 1997, the physicist Freeman Dyson thought of a clever way to keep a civilization alive even as the universe dies out: hibernation. Dyson argued that creatures in a dying galaxy could set up machines that collect energy (and shed entropy) while the creatures sleep, unconscious, in a state of hibernation. When the machines have collected enough energy and brought the civilization's immediate environment sufficiently out of equilibrium, the creatures awaken. They live off the collected energy for a while, discard their entropy into their environment, and process and repair the damage that Nature has done to their stored information. As the energy is used up and their environment once more reaches equilibrium, they go back to sleep until the machines make the conditions right for them to awaken once more.

But in 1999, Lawrence Krauss, a physicist at Case Western Reserve University, showed that the hibernation scheme ultimately had to fail. As the universe reaches equilibrium, the energy-gathering–entropy-shedding machines take longer and longer to do their task—to gather the required energy and shed the required entropy to awaken the creatures. The periods of hibernation must get dramatically longer and the periods of consciousness must get dramatically shorter as the universe expands and dies. As the universe reaches equilibrium, after a certain point the machines can chug away forever and they will not ever collect enough energy and shed enough entropy to give the civilization even another second of consciousness. The information processing stops forever; the information so carefully stored by the civilization over the millennia slowly dissipates into the environment, and equilibrium and entropy bring darkness to the last living civilization. Life becomes extinct.

It's a dark picture, but physicists have come to the same conclusion in a different way. Our universe (or multiverse) is constantly churning. Information passes back and forth and the environment (conscious or otherwise) processes it and dissipates it. In a sense, the universe as a whole is behaving like a giant information processor—a computer.

So, if the universe can, even in the abstract, be considered a computer, how many operations has it performed? How many operations can it perform in the future? Thanks to the laws of information, scientists have answered both questions.

In 2001, Seth Lloyd, the physicist who discovered that a black hole would be the ultimate laptop, used similar logic to figure out how many computations the visible universe, our Hubble bubble, could have performed since the big bang. Through the energy-time relationship, the amount of matter and energy in the universe determines how fast those computations can be performed—yielding an enormous 10^{120} operations from the beginning of time until today. In 2004, Krauss did the other side of the calculation—the amount of computations that can be done in the future. In an ever-expanding universe, that number is finite in our Hubble bubble, and it appears to be just a tad more than 10^{120} operations—almost precisely the same as the maximum number of operations that could have been done in the past. The number 10^{120} is huge—but it's finite. There are a limited number of information-processing operations that our Hubble bubble has left in it. Inasmuch as life relies upon information processing, it, too, must be finite. Life cannot go on forever. It has at most 10^{120} operations left, and then all life in the visible universe will go extinct. The information stored and preserved by those living creatures will then be irreversibly dissipated. Though the information is never truly destroyed, it will be scattered, useless, throughout the dark, lifeless cosmos.

This is the ultimate irony of the laws of information. Physicists are using information to figure out the most profound questions of the universe. What are the ultimate laws of physics? What accounts for the weirdness of relativity and quantum mechanics? What lies at the center of a black hole? Is ours the only universe or are there others? What is the structure of the universe? What is life? Using the tools of information theory, scientists are beginning to get answers to all of these questions. But at the same time, those tools of information theory have revealed our ultimate fate. We will die, as will all the answers we have to these questions—all the information our civilization has gathered. Life must end, and with it will end all consciousness, all ability to understand the universe. Using information, we may find the ultimate answers, yet those answers will be rendered worthless by the laws of information.

This precious information that may well illuminate the darkest mysteries about the universe carries in it the seeds of its own destruction.

THE LOGARITHM

The logarithm is the opposite of exponentiation, just as division is the opposite of multiplication.

To undo a multiplication by 6, you divide by 6: $5 \times 6 = 30$, and $30 \div 6 = 5$.

To undo an exponentiation where 6 is raised to a certain power, you take the logarithm, base 6. That is: $6^5 = 7776$, and $\log_6 7776 = 5$, where $\log_6$ represents the logarithm base 6.

You will seldom see the base written explicitly, and this can be a source of confusion, because *log* can mean different things in different contexts. Most of the time, log means the logarithm base 10. So, usually $\log 1000 = 3$, because $\log_{10} 10^3 = 3$.

However, that's not a universal convention. Many computer scientists think in terms of binary numbers, and to them it's more useful to have log mean log base 2. To these computer scientists, $\log 1000$ isn't $\log_{10} 1000$; it's $\log_2 1000$, which is a bit less than 10. And to many mathematicians, it's more natural to think in terms of a number between 2 and 3 known as e; to them, $\log 1000$ is really $\log_e 1000$, which is roughly 7.

(Nonmathematicians often use the symbol "ln" to represent "$\log_e$," but this is not universal among mathematicians.)

You'd think that this would cause a lot of problems, but in fact, it really doesn't make too much of a difference *what* base the logarithm is. They are all so closely related that in many equations the base can be left ambiguous.

For example, in the Boltzmann equation, $S = k \log W$, it doesn't matter whether the log is base 2, base 10, base e, or base 42. Your choice of base gets absorbed into the k. For example, let's assume that the above equation refers to the logarithm base 10. It turns out that

$$S = k \log_{10} W = k\,(\log_{10} 42)(\log_{42} W) = k' \log_{42} W$$

where k' is our new constant—k multiplied by $\log_{10} 42$. The equation looks exactly the same in base 42 as it does in base 10: $S = k' \log W$, but this time, log refers to $\log_{42}$ rather than $\log_{10}$. You can completely ignore the base, and the equation looks precisely the same.

For this reason, I have used the symbol "log" to refer to the logarithm without specifying which base. In the Boltzmann equation, it tends to refer to log base 10 or log base e, depending on which value of the constant is used; in Shannon entropy, it is log base 2; and later on in the book, when it comes to erasure, energy, entropy, and computing, we're back to log base e.

Since the differences to the equations in question are entirely cosmetic, for the sake of clarity I have consistently omitted the base when using the logarithm.

ENTROPY AND INFORMATION

In this book, the equation for entropy (symbolized by the letter S) has been given in three different forms. Though the forms look somewhat different, in fact they are all the same.

The first equation for entropy is Boltzmann's: $S = k \log W$, where W is the number of ways that the system can have the state whose entropy we're calculating.

The second equation for entropy is one that I derive for a specific system—tossing marbles in a box—which is $S = k \log p$, where p is the probability of a given configuration of marbles in the box. Actually, I say S is a *function* of $k \log p$—more on this in a moment.

The third equation for entropy is Shannon's, which I didn't give explicitly in the main text. For the case we are concerned with, $S = -\sum p_i \log p_i$, where each p_i represents the probability of any particular message in the collection of possible messages a source could have sent you, and the Greek letter sigma, $\sum$, represents the sum of all of these terms. (Incidentally, the p_is can represent possible symbols rather than possible messages; the result comes out the same, but the math for that example would be a little more complicated, since it would require using conditional probabilities, which require a longer chain of argument.)

Let's use all three of these equations to analyze a system. Say, for example, someone drops four identical marbles into a box and then walks away; there is an equal chance that any given marble falls into the left side or the right. Later, when we approach the box and look in, we see that two marbles are on the left and two are on the right. What's the entropy of the system?

According to Boltzmann's equation, $S = k \log W$, W represents the number of ways we can get the state in question, namely two marbles on each side of the box. There are, in fact, six ways (1 and 2 land on the right, or 1 and 3 land there, or 1 and 4, or 2 and 3, or 2 and 4, or 3 and 4). So, the entropy, $S = k \log 6$.

According to the marbles-in-a-box derivation, S is a function of $k \log p$. More specifically, $S = k \log p + k \log N$, where N is the number of ways that distinguishable marbles can be arranged in the box; in this particular example, N is 16. (The $k \log N$ term serves simply to keep the entropy from going negative; dropping it would make little difference.)

The probability of having two marbles on each side of the box is 3/8, as shown in the table in chapter 2, so $S = k \log(3/8) + k \log 16$. But 3/8 is the same thing as 6/16, and $\log 6/16$ is the same thing as $\log 6 - \log 16$. Thus, we find that the marbles-in-a-box formulation is $S = k \log 6 - k \log 16 + k \log 16$, which, of course, is just $k \log 6$.

The Shannon equation deals with messages rather than marbles in boxes, but we can easily convert between one and the other. Let's say that a **1** represents a marble dropping into the right side of the box and a **0** represents a marble falling into the left side. When we drop marbles into the box, we can write the result in a message of bits: **1100** means that in a sequence of four marbles tossed in the box, marbles 1 and 2 wound up on the right and marbles 3 and 4 wound up on the left. Looking at the box, we see two marbles on each side, so we know that the system must have received one of the six following messages: **1100, 1010, 1001, 0110, 0101, 0011.**

When we look at the box, we don't know *which* of these messages was received; we don't know which particular marbles are on the right and which are on the left because they all look the same. But we know that one of those six messages was received. Because of the way the system was set up—a 50:50 chance of a marble's falling on the left or the right—we know that each of those messages is equally probable. So, given our knowledge of the system, we can see that each message has a probability of 1/6 associated with it. This means that the expression $-\sum p_i \log p_i$ has six terms—one for each possible message—and each p_i, each probability in the expression, is 1/6. So,

$$
\begin{aligned}
S &= -\sum p_i \log p_i \\
&= -[(1/6)\log(1/6) + (1/6)\log(1/6) + (1/6)\log(1/6) \\
&\quad + (1/6)\log(1/6) + (1/6)\log(1/6) + (1/6)\log(1/6)] \\
&= -6[(1/6)\log(1/6)] \\
&= -\log(1/6).
\end{aligned}
$$

But $-\log(1/6)$ is the same thing as $\log 6$, so we get $S = \log 6$. Where did the k go? Well, the log here is base 2, which is not the same base we used above. The k disappeared because, as we saw in appendix A, changing the base of a logarithm in this case merely changes the look of the constant k; in base 2, and using units that are slightly different from the ones used in Boltzmann's equation, our new k equals 1.

OK—so Shannon entropy is the same as Boltzmann's thermodynamic entropy is the same as the marbles-in-a-box entropy. How does entropy relate to information? This is a complicated question, and it is a major source of confusion.

The entropy of a message source is equivalent to the amount of information it can send in any given message. Say we've got a source that produces strings of eight bits; each eight-bit message is equally probable. It has an entropy of eight bits, and each message can carry eight bits of information. A generic message from this source would look something like: **10110101**. Most likely, it would seem pretty random.

On the other hand, a source that produces strings of eight bits where, say, only the messages **00000000** and **11111111** are possible, has a much lower entropy: only one bit. Each message can only carry one bit of information. A generic message from this source would look like either **00000000** or **11111111**—not very "random-seeming" at all. It's a general principle; the more "random-seeming" the message you have received, the higher the entropy (in general) of the source of the message, and the more information (in general) the message can contain.

But with information, you can look at it from the receiver's point of view rather than the sender's, and the situation is, in a sense, reversed. And this can be incredibly confusing.

Remember, information is the answer to a question of some sort: information reduces your uncertainty about which of the possible answers is the correct one. Back to the four-marble system. Say that you want to know the answer to the question, Where did marble 1 land? If two marbles are on the right and two marbles are on the left, we have absolutely no information whatsoever about where marble 1 is; there's a 50:50 chance it's on either side. If there are three marbles on the right and one marble on the left, we have a little more certainty, a little more information about the answer to the question; marble 1 is probably on the right. There is a 75 percent chance that it is one of three marbles on the right, versus a 25 percent chance it is the marble on the left. And if all four marbles are on the right side, we have absolute certainty. There's a 100 percent chance that marble 1 landed on the right. This time, the *lower* the entropy of the system, the more information we have about where marble 1 landed.

It gets worse. Using our **0** and **1** code as before, "random-seeming" streams like **0110** and **1100**, where two balls are on each side, mean *more* uncertainty than nonrandom streams like **1111** and **0000**, where we know, for certain, which side marble 1 has landed on: the *less* random the stream of symbols, the more information we have about the position of marble 1. This appears to be just the opposite of our previous analysis.

However, it makes sense if you think about it carefully. Information is flowing from the sender to the recipient of a message, and each has a different role in the transaction. Entropy is a measure of ambiguity, unpredictability, and uncertainty, and it is really good for a source of a message to have a high entropy. It means that the source is unpredictable, and you don't know what a message from that source is going to say ahead of time. (If you always know what the message will say, it wouldn't give you any information, would it?) But once the recipient receives the message, that message, if it contains lots of information, should reduce the uncertainty about the answer to a question. The more entropy, the more uncertainty you have about an answer, the less information you must have received.

Sometimes you will hear people say entropy is the same thing as information; sometimes you will hear people say that information is *negative* entropy or *negentropy*. The difference arises because people are accustomed to analyzing different things. Some are looking at the sender and the unpredictability of a potential message, and some are looking at the receiver and the uncertainties about the answer to a question. In truth, both are looking at the same thing: sender and receiver are two sides of the same coin.

SELECT BIBLIOGRAPHY

Albrecht, Andreas. "Cosmic Inflation and the Arrow of Time." In arXiv.org e-Print archive (www.arxiv.org), astro-ph/0210527, 24 October 2002.

Associated Press. "DNA Links Teacher to 9,000-Year-Old Skeleton." CNN, 7 May 1997. Human Origins Web site. www.versiontech.com/origins/news/news_article.asp?news_id=13

Bacciagaluppi, Guido. "The Role of Decoherence in Quantum Theory." *The Stanford Encyclopedia of Philosophy,* Edward N. Zalta, ed. plato.stanford.edu/entries/qm-decoherence/

Baez, John. "The Quantum of Area?" *Nature* 421 (2003): 702.

Bejerano, Gill, Michael Pheasant, Igor Makunin, Stuart Stephen, W. James Kent, John S. Muttick, and David Haussler. "Ultraconserved Elements in the Human Genome." *ScienceExpress,* 6 May 2004. www.sciencemag.org/cgi/rapidpdf/1098119v1.pdf

Bekenstein, Jacob D. "Information in the Holographic Universe." *Scientific American,* August 2003, 48.

Blanton, John. "The EPR Paradox and Bell's Inequality Principle." University of California, Riverside, Department of Mathematics Web site. math.ucr.edu/home/baez/physics/Quantum/bells_inequality.html

Bohinski, Robert C. *Modern Concepts in Biochemistry.* Boston: Allyn and Bacon, 1987.

Boyce, Nell. "Dangerous Liaison." *New Scientist,* 19/26 December 1998, 21.

Bradman, Neil, and Mark Thomas. "Why Y?" *Science Spectra,* no. 14, 1998. www.ucl.ac.uk/tcga/ScienceSpectra-pages/SciSpect-14-98.html

Brezger, B., L. Hackermüller, S. Uttenthaler, J. Petschinka, M. Arndt, and A. Zeilinger. "Matter-Wave Interferometer for Large Molecules." In arXiv.org e-Print archive (www.arxiv.org), quant-ph/0202158, 26 February 2002.

Brookes, Martin. "Apocalypse Then." *New Scientist,* 14 August 1999, 32.

Brukner, Časlav, Markus Aspelmeyer, and Anton Zeilinger. "Complementarity and Information in 'Delayed-Choice for Entanglement Swapping.'" In arXiv.org e-Print archive (www.arxiv.org), quant-ph/0405036, 7 May 2004.

Brukner, Časlav, and Anton Zeilinger. "Information and Fundamental Elements of the Structure of Quantum Theory." In arXiv.org e-Print archive (www.arxiv.org), quant-ph/0212084, 13 December 2002.

———. "Operationally Invariant Information in Quantum Measurements." In arXiv.org e-Print archive (www.arxiv.org), quant-ph/0005084, 19 May 2000.

Budnik, Paul. "Measurement in Quantum Mechanics FAQ." Mountain Math Software Web site. www.mtnmath.com/faq/meas-qm.html

Calderbank, A. R., and Peter W. Shor. "Good Quantum Error-Correcting Codes Exist." In arXiv.org e-Print archive (www.arxiv.org), quant-ph/9512032, 16 April 1996.

Carberry, D. M., J. C. Reid, G. M. Wang, E. M. Sevick, Debra J. Searles, and Denis J. Evans. "Fluctuations and Irreversibility: An Experimental Demonstration of a Second-Law-Like Theorem Using a Colloidal Particle Held in an Optical Trap." *Physical Review Letters* 92 (2004): art. no. 140601.

The Catholic Encyclopedia. www.newadvent.org/cathen

Cavalli-Sforza, Luigi Luca. *Genes, Peoples, and Languages.* Mark Seislstad, trans. New York: North Point Press, 2000.

Cavalli-Sforza, Luigi Luca, Paolo Menozzi, and Alberto Piazza. *The History and Geography of Human Genes.* Princeton: Princeton University Press, 1994.

"Central Bureau—Interception and Cryptanalyzing of Japanese Intelligence." Web site. home.st.net.au/~dunn/sigint/cbi.htm

Chiao, Raymond Y., Paul G. Kwiat, and Aephraim M. Steinberg. "Quantum Nonlocality in Two-Photon Experiments at Berkeley." In arXiv.org e-Print archive (www.arxiv.org), quant-ph/9501016, 18 January 1995.

Cornish, Neil J., Daniel N. Spergel, Glenn D. Starkmann, and Eiichiro Komatsu. "Constraining the Topology of the Universe." In arXiv.org e-Print archive (www.arxiv.org), astro-ph/0310233, 8 October 2003.

Dawkins, Richard. *The Extended Phenotype.* New York: Oxford University Press, 1999.

———. *The Selfish Gene.* New York: Oxford University Press, 1989.

de Mendoza, Diego Hurtado, and Ricardo Braginski. "Y Chromosomes Point to Native American Adam." *Science* 283 (1999): 1439.

Derix, Martijn, and Jan Pieter van der Schaar. "Black Hole Physics," from "Stringy Black Holes." www-th.phys.rug.nl/~schaar/htmlreport/node8.html (linked from Jan Pieter van der Schaar's home page: vangers.home.cern.ch/vanders).

de Ruyter van Steveninck, Rob, Alexander Borst, and William Bialek. "Real Time Encoding of Motion: Answerable Questions and Questionable Answers from the Fly's Visual System." In arXiv.org e-Print archive (www.arxiv.org), physics/0004060, 25 April 2000.

Deutsch, David. "The Structure of the Multiverse." In arXiv.org e-Print archive (www.arxiv.org), quant-ph/0104033, 6 April 2001.

Deutsch, David, and Patrick Hayden. "Information Flow in Entangled Quantum Systems." In arXiv.org e-Print archive (www.arxiv.org), quant-ph/9906007, 1 June 1999.

Dokholyan, Nikolay V., Sergey V. Buldyrev, Shlomo Havlin, and H. Eugene Stanley. "Distribution of Base Pair Repeats in Coding and Noncoding DNA Sequences." *Physical Review Letters* 79 (1997): 5182.

Einstein, Albert. *Relativity: The Special and the General Theory.* New York: Crown, 1961.

Faulhammer, Dirk, Anthony R. Cukras, Richard J. Lipton, and Laura F. Landweber. "Molecular Computation: RNA Solutions to Chess Problems." *Proceedings of the National Academy of Sciences* 97 (2000): 1385.

Feynman, Richard, Robert B. Leighton, and Matthew Sands. *The Feynman Lectures on Physics.* 3 vols. Reading, Mass.: Addison-Wesley, 1989.

Fondation Odier de Psycho-Physique. Bulletin no. 4. Geneva, Switzerland: 2002.

Garriga, J., V. F. Mukhanov, K. D. Olum, and A. Vilenkin. "Eternal Inflation, Black Holes, and the Future of Civilizations." In arXiv.org e-Print archive (www.arxiv.org), astro-ph/9909143, 16 May 2000.

Garriga, Jaume, and Alexander Vilenkin. "In Defence of the 'Tunneling' Wave Function of the Universe." In arXiv.org e-Print archive (www.arxiv.org), gr-qc/9609067, 30 September 1996.

———. "Many Worlds in One." In arXiv.org e-Print archive (www.arxiv.org), gr-qc/0102010, 2 May 2001.

"The German Enigma Cipher Machine—History of Solving." Web site. www.enigmahistory.org/chronology.html

Gershenfeld, Neil A., and Isaac L. Chuang. "Bulk Spin-Resonance Quantum Computation." *Science* 275 (1997): 350.

Gettemy, Charles. "The Midnight Ride of April 18, 1775." Chap. 3 in *The True Story of Paul Revere.* Archiving Early America Web site. earlyamerica.com/lives/revere/chapt3/

Gilchrist, A., Kae Nemeto, W. J. Munroe, T. C. Ralph, S. Glancey, Samuel L. Braunstein, and G. J. Milburn. "Schrödinger Cats and Their Power for Quantum Information Processing." In arXiv.org e-Print archive (www.arxiv.org), quant-ph/0312194, 24 December 2003.

Griffiths, Robert B. "Consistent Histories and Quantum Reasoning." In arXiv.org e-Print archive (www.arxiv.org), quant-ph/9606005, 4 June 1996.

Hackermüller, Lucia, Klaus Hornberger, Björn Brezger, Anton Zeilinger, and Marcus Arndt. "Decoherence of Matter Waves by Thermal Emission of Radiation." *Nature* 427 (2004): 711.

Hackermüller, Lucia, Stefan Uttenthaler, Klaus Hornberger, Elisabeth Reiger, Björn Brezger, Anton Zeilinger, and Markus Arndt. "The Wave Nature of Biomolecules and Fluorofullerenes." In arXiv.org e-Print archive (www.arxiv.org), quant-ph/0309016, 1 September 2003.

Harpending, Henry C., Mark A. Batzer, Michael Gurven, Lynn B. Jorde, Alan R. Rogers, and Stephen T. Sherry. "Genetic Traces of Ancient Demography." *Proceedings of the National Academy of Sciences* 95 (1998): 1961.

Harrison, David M. "Black Hole Thermodynamics." UPSCALE Web site, Department of Physics, University of Toronto. www.upscale.utoronto.ca/GeneralInterest/Harrison/BlackHoleThermo/BlackHoleThermo.html

Hawking, Stephen. *A Brief History of Time: From the Big Bang to Black Holes.* New York: Bantam, 1988.

Heisenberg, Werner. *Physics and Philosophy: The Revolution in Modern Science.* New York: Harper & Row, 1958.

Herodotus. *The Histories.* Aubrey de Selincourt, trans. London: Penguin, 1996.

Hill, Emmeline W., Mark A. Jobling, and Daniel G. Bradley. "Y-chromosome Variation and Irish Origins." *Nature* 404 (2000): 351.

Hodges, Andrew. *Alan Turing: The Enigma.* New York: Walker, 2000.

Holevo, A. S. "Coding Theorems for Quantum Communication Channels." In arXiv.org e-Print archive (www.arxiv.org), quant-ph/9708046, 27 August 1997.

———. "Remarks on the Classical Capacity of a Quantum Channel." In arXiv.org e-Print archive (www.arxiv.org), quant-ph/0212025, 4 December 2002.

Hornberger, Klaus, Stefan Uttenthaler, Björn Brezger, Lucia Hackermüller, Markus Arndt, and Anton Zeilinger. "Collisional Decoherence Observed in Matter Wave Interferometry." In arXiv.org e-Print archive (www.arxiv.org), quant-ph/0303093, 14 March 2003.

Imperial War Museum. "The Battle of the Atlantic." www.iwm.org.uk/online/atlantic/dec41dec42.htm

"ISBN." eNSYNC Solutions Web site. www.ensyncsolutions.com/isbn.htm

Johnson, Welin E., and John M. Coffin. "Constructing Primate Phylogenies from Ancient Retrovirus Sequences." *Proceedings of the National Academy of Sciences* 96 (1999): 10254.

Kiefer, Claus, and Erich Joos. "Decoherence: Concepts and Examples." In arXiv.org e-Print archive (www.arxiv.org), quant-ph/9803052, 19 March 1998.

Knill, E., R. Laflamme, R. Martinez, and C.-H. Tseng. "An Algorithmic Benchmark for Quantum Information Processing." *Nature* 404 (2000): 368.

Kofman, A. G., and G. Kurizki. "Acceleration of Quantum Decay Processes by Frequent Observations." *Nature* 405 (2000): 546.

Kornguth, Steve. "Brain Demystified." www.lifesci.utexas.edu/courses/ brain/ Steve'sLectures/neuroimmunol/MetabolismImaging.html (no longer available).

Krauss, Lawrence M., and Glenn D. Starkman. "Life, The Universe, and Nothing: Life and Death in an Ever-Expanding Universe." In arXiv.org e-Print archive (www.arxiv.org), astro-ph/9902189, 12 February 1999.

———. "Universal Limits on Computation." In arXiv.org e-Print archive (www.arxiv.org), astro-ph/0404510, 26 April 2004.

Kraytsberg, Yevgenya, Marianne Schwartz, Timothy A. Brown, Konstantin Ebralidse, Wolfram S. Kunz, David A. Clayton, John Vissing, and Konstantin Khrapko. "Recombination of Human Mitochondrial DNA." *Science* 304 (2004): 981.

Kukral, L. C. "Death of Yamamoto due to 'Magic.'" Navy Office of Information Web site. www.chinfo.navy.mil/navpalib/wwii/facts/yamadies.txt

Kunzig, Robert, and Shanti Menon. "Not Our Mom: Neanderthal DNA Suggests No Relation to Humans." *Discover,* January 1998, 32.

Lamoreaux, S. K. "Demonstration of the Casimir Force in the 0.6 to 6μm Range." *Physical Review Letters* 77 (1997): 5.

Leff, Harvey S., and Andrew F. Rex, eds. *Maxwell's Demon 2: Entropy, Classical and Quantum Information, Computing.* 2nd ed. Philadelphia: Institute of Physics Publishing, 2003.

Lindley, David. *Boltzmann's Atom.* New York: Free Press, 2001.

———. *Degrees Kelvin.* Washington, D.C.: Joseph Henry Press, 2004.

Lloyd, Seth. "Computational Capacity of the Universe." In arXiv.org e-Print archive (www.arxiv.org), quant-ph/0110141, 24 October 2001.

———. "Ultimate Physical Limits to Computation." *Nature* 406 (2000): 1047.

———. "Universe as Quantum Computer." In arXiv.org e-Print archive (www.arxiv.org), quant-ph/9912088, 17 December 1999.

Macrae, Norman. *John von Neumann: The Scientific Genius Who Pioneered the Modern Computer, Game Theory, Nuclear Deterrence, and Much More.* New York: Pantheon, 1992.

Marangos, Jon. "Faster Than a Speeding Photon." *Nature* 406 (2000), 243.

Marcikic, I., H. de Riedmatten, W. Tittel, H. Zbinden, and N. Gisin. "Long-Distance Teleportation of Qubits at Telecommunication Wavelengths." *Nature* 421 (2003): 509.

Marshall, William, Christoph Simon, Roger Penrose, and Dik Bouwmeester. "Towards Quantum Superpositions of a Mirror." In arXiv.org e-Print archive (www.arxiv.org), quant-ph/0210001, 30 September 2002.

Mermin, N. David. "Could Feynman Have Said This?" *Physics Today,* 57 (2004): 10.

Miller, A. Ray. *The Cryptographic Mathematics of Enigma.* Fort George G. Meade, Maryland: The Center for Cryptologic History, 2002.

Milonni, Peter. "A Watched Pot Boils Quicker." *Nature* 405 (2000): 525.

Mitchell, Morgan W., and Raymond Y. Chiao. "Causality and Negative Group Delays in a Simple Bandpass Amplifier." *American Journal of Physics* 66 (1998): 14.

Monroe, C., D. M. Meekhof, B. E. King, and D. J. Wineland. "A 'Schrödinger Cat' Superposition State of an Atom." *Science* 272 (1996): 1131.

Mugnai, D., A. Ramfagni, and R. Ruggeri. "Observation of Superluminal Behaviors in Wave Propagation." *Physical Review Letters* 84 (2000): 4830.

The Museum of Science & Industry in Manchester. "Joule & Energy." www.msim.org.uk/joule/intro.htm

Naughton, John. "The Juggling Unicyclist Who Changed Our Lives." *The Observer,* 4 March 2001. observer.guardian.co.uk/business/story/0,6903,446009,00.html

Naval Historical Center, Department of the Navy. "Battle of Midway: 4–7 June 1942." www.history.navy.mil/faqs/faq81-1.htm

Nemenman, Ilya, William Bialek, and Rob de Ruyter van Stevenick. "Entropy and Information in Neural Spike Trains: Progress on the Sampling Problem." In arXiv.org e-Print archive (www.arxiv.org), physics/0306063, 12 March 2004.

Ollivier, Harold, David Poulin, and Wojciech H. Zurek. "Emergence of Objective Properties from Subjective Quantum States: Environment as a Witness." In arXiv.org e-Print archive (www.arxiv.org), quant-ph/0307229, 30 July 2003.

Owens, Kelly, and Mary-Claire King. "Genomic Views of Human History." *Science* 286 (2000): 451.

Pais, Abraham. *Subtle Is the Lord . . . : The Science and the Life of Albert Einstein.* Oxford: Oxford University Press, 1982.

Park, Yousin. "Entropy and Information." Physics & Astronomy @ Johns Hopkins Web site. www.pha.jhu.edu/~xverver/seminar2/seminar2.html

Pati, Arun, and Samuel Braunstein. "Quantum Deleting and Signalling." In arXiv.org e-Print archive (www.arxiv.org), quant-ph/0305145, 23 May 2003.

———. "Quantum Mechanical Universal Constructor." In arXiv.org e-Print archive (www.arxiv.org), quant-ph/0303124, 19 March 2003.

———. "Quantum No-Deleting Principle and Some of Its Implications." In arXiv.org e-Print archive (www.arxiv.org), quant-ph/0007121, 31 July 2000.

The Paul Revere House. "The Midnight Ride." www.paulreverehouse.org/ride/

Pennisi, Elizabeth. "Viral Stowaway." *ScienceNOW,* 1 March 1999. sciencenow.sciencemag.org/cgi/content/full/1999/301/1

Peres, Asher. "How the No-Cloning Theorem Got Its Name." In arXiv.org e-Print archive (www.arxiv.org), quant-ph/0205076, 14 May 2002.

Pincus, Steve, and Rudolf Kalman. "Not All (Possibly) 'Random' Sequences Are Created Equal." *Proceedings of the National Academy of Sciences* 94 (1997): 3513.

Preskill, John. "Black Hole Information Bet." Caltech Particle Theory Group Web site. www.theory.caltech.edu/people/preskill/info_bet.html

———. "Reliable Quantum Computers." In arXiv.org e-Print archive (www.arxiv.org), quant-ph/9705031, 1 June 1997.

Price, Michael Clive. "The Everett FAQ." HEDWEB site. www.hedweb.com/manworld.htm

Russell, Jerry C. "ULTRA and the Campaign Against the U-boats in World War II." Ibiblio archive. www.ibiblio.org/pha/ultra/navy-1.html

Schneidman, Elad, William Bialek, and Michael J. Berry II. "An Information Theoretic Approach to the Functional Classification of Neurons." In arXiv.org e-Print archive (www.arxiv.org), physics/0212114, 31 December 2002.

Schrödinger, Erwin. *What Is Life?* Cambridge: Cambridge University Press, 1967.

Sears, Francis W., Mark W. Zemansky, and Hugh D. Young. *College Physics,* 6th ed. Reading, Mass.: Addison-Wesley, 1985.

Seife, Charles. "Alice Beams Up 'Entangled' Photon." *New Scientist,* 12 October 1997, 20.

———. *Alpha and Omega: The Search for the Beginning and End of the Universe.* New York: Viking, 2003.

———. "At Canada's Perimeter Institute, 'Waterloo' Means 'Shangri-La.'" *Science* 302 (2003): 1650.

———. "Big, Hot Molecules Bridge the Gap Between Normal and Surreal." *Science* 303 (2004): 1119.

———. "Cold Numbers Unmake the Quantum Mind." *Science* 287 (2000): 791.

———. "Crystal Stops Light in Its Tracks." *Science* 295 (2002): 255.

———. "Flaw Found in a Quantum Code." *Science* 276 (1997): 1034.

———. "Furtive Glances Trigger Radioactive Decay." *Science* 288 (2000): 1564.

———. "In Clone Wars, Quantum Computers Need Not Apply." *Science* 300 (2003): 884.

———. "Light Speed Boosted Beyond the Limit." *ScienceNOW,* 21 July 2000. sciencenow.sciencemag.org/cgi/content/full/2000/721/4

———. "Messages Fly No Faster Than Light." *ScienceNOW,* 15 October 2003. sciencenow.sciencemag.org/cgi/content/full/2003/1015/2

———. "Microscale Weirdness Expands Its Turf." *Science* 292 (2001): 1471.

———. "More Than We Need to Know." *Washington Post,* 9 November 2001, A37.

———. "Muon Experiment Challenges Reigning Model of Particles." *Science* 291 (2001): 958

———. "Perimeter's Threefold Way." *Science* 302 (2003): 1651.

———. "The Quandary of Quantum Information." *Science* 293 (2001): 2026.

———. "Quantum Experiment Asks 'How Big Is Big?'" *Science* 298 (2002): 342.

———. "Quantum Leap." *New Scientist,* 18 April 1998, 10.

———. "Relativity Goes Where Einstein Feared to Tread." *Science* 299 (2003): 185.

———. "RNA Works Out Knight Moves." *Science* 287 (2000): 1182.

———. "Souped Up Pulses Top Light Speed." *ScienceNOW,* 1 June 2000. sciencenow.sciencemag.org/cgi/content/full/2000/601/1

———. "'Spooky Action' Passes a Relativistic Test." *Science* 287 (2000): 1909.

———. "Spooky Twins Survive Einsteinian Torture." *Science* 294 (2001): 1265.

———. "The Subtle Pull of Emptiness." *Science* 275 (1997): 158.

———. "'Ultimate PC' Would Be a Hot Little Number." *Science* 289 (2000): 1447.

———. *Zero: The Biography of a Dangerous Idea.* New York: Viking, 2000.

Shannon, Claude E., and Warren Weaver. *The Mathematical Theory of Communication.* Urbana: University of Illinois Press, 1998.

Siegfried, Tom. *The Bit and the Pendulum: From Quantum Computing to M Theory—The New Physics of Information.* New York: John Wiley, 1999.

Singh, Simon. *The Code Book: The Evolution of Secrecy from Mary Queen of Scots to Quantum Cryptography.* New York: Doubleday, 1999.

Sloane, N. J. A., and A. D. Wyner. "Biography of Claude Elwood Shannon." AT&T Labs–Research Web site. www.research.att.com/~njas/doc/shannonbio.html

Stefanov, Andre, Hugo Zbinden, Antoine Suarez, and Nicolas Gisin. "Quantum Entanglement with Acousto-Optic Modulators: 2-Photon Beatings and Bell Experiments with Moving Beamsplitters." In arXiv.org e-Print archive (www.arxiv.org), quant-ph/0210015, 2 October 2002.

Steinberg, A. M., P. G. Kwiat, and R. Y. Chiao. "Measurement of the Single-Photon Tunneling Time." *Physical Review Letters* 71 (1993): 708.

Stenner, Michael D., Daniel J. Gauthier, and Mark A. Neifeld. "The Speed of Information in a 'Fast-Light' Optical Medium." *Nature* 425 (2003): 695.

Strauss, Evelyn. "Can Mitochondrial Clocks Keep Time?" *Science* 283 (1999): 1435.

Strong, S. P., Roland Koberle, Rob R. de Ruyter van Steveninck, and William Bialek. "Entropy and Information in Neural Spike Trains." *Physical Review Letters* 80 (1998): 197.

Suarez, Antoine. "Is There a Real Time Ordering Behind the Nonlocal Correlations?" In arXiv.org e-Print archive (www.arxiv.org), quant-ph/0110124, 20 October 2001.

Teahan, W. J., and John G. Cleary. "The Entropy of English Using PPM-Based Models." The University of Waikato, Department of Computer Science Web site. www.cs.waikato.ac.nz/~ml/publications/1996/Teahan-Cleary-entropy96.pdf

Tegmark, Max. "Importance of Decoherence in Brain Processes." *Physical Review E* 61 (2000): 4194.

———. "The Interpretation of Quantum Mechanics: Many Worlds or Many Words?" In arXiv.org e-Print archive (www.arxiv.org), quant-ph/9709032, 15 September 1997.

———. "Parallel Universes." In arXiv.org e-Print archive (www.arxiv.org), astro-ph/0302131, 7 February 2003.

Tegmark, Max, and John Archibald Wheeler. "100 Years of the Quantum." In arXiv.org e-Print archive (www.arxiv.org), quant-ph/0101077, 17 January 2001.

't Hooft, Gerard. *In Search of the Ultimate Building Blocks.* Cambridge: Cambridge University Press, 1997.

Thorne, Kip S. *Black Holes and Time Warps: Einstein's Outrageous Legacy.* New York: W. W. Norton, 1994.

Tribus, Myron, and Edward McIrvine. "Energy and Information." *Scientific American,* August 1971, 179.

"UCLA Brain Injury Research Center Project Grants." UCLA Neurosurgery Web site. neurosurgery.ucla.edu/Programs/BrainInjury/BIRC_project.html

Voss, David. "'New Physics' Finds a Haven at the Patent Office." *Science* 284 (1999): 1252.

Wang, G. M., E. M. Sevick, Emil Mittag, Debra J. Searles, and Denis J. Evans. "Experimental Demonstration of Violations of the Second Law of Thermodynamics for Small Systems and Short Time Scales." *Physical Review Letters* 89 (2002): art. no. 050601.

Wang, L. J., A. Kuzmich, and A. Dogariu. "Demonstration of Gain-Assisted Superluminal Light Propagation." Dr. Lijun Wang's home page. external.nj. nec.com/homepages/lwan/demo.htm

———. "Gain-Assisted Superluminal Light Propagation." *Nature* 406 (2000): 277.

Weadon, Patrick D. "AF Is Short of Water," from "The Battle of Midway." National Security Agency Web site. www.nsa.gov/publications/publi00023.cfm

Weisstein, Eric. Eric Weisstein's World of Science. scienceworld.wolfram.com

Whitaker, Andrew. *Einstein, Bohr and the Quantum Dilemma.* New York: Cambridge University Press, 1996.

Zeh, H. D. "Basic Concepts and Their Interpretation." Preliminary version of chap. 2, *Decoherence and the Appearance of a Classical World in Quantum Theory,* 2nd ed., by E. Joos, H. D. Zeh, C. Kiefer, D. J. W. Giulini, J. Kupsch, and I.-O. Stamatescu (Springer, 2003). www.rzuser.uni-heidelberg.de/~as3/Decoh2.pdf (linked from H. Dieter Zeh's home page: www.zeh-hd.de).

———. "The Meaning of Decoherence." In arXiv.org e-Print archive (www.arxiv.org), quant-ph/9905004, 29 June 1999.

———. *The Physical Basis of the Direction of Time.* Berlin: Springer-Verlag, 2001.

———. "The Wave Function: It or Bit?" In arXiv.org e-Print archive (www.arxiv.org), quant-ph/0204088, 2 June 2002.

Zurek, Wojciech Hubert. "Decoherence and the Transition from Quantum to Classical—Revisited." *Los Alamos Science* no. 27 (2002): 2.

———. "Quantum Darwinism and Envariance." In arXiv.org e-Print archive (www.arxiv.org), quant-ph/03080163, 28 August 2003.

———. "Quantum Discord and Maxwell's Demons." In arXiv.org e-Print archive (www.arxiv.org), quant-ph/0301127, 23 January 2003.

ACKNOWLEDGMENTS

This book is the product of years of discussion and research, and it would be impossible to thank all the people who were involved. Dozens of physicists, quantum theorists, cosmologists, astronomers, biologists, cryptographers, and other scientists have been extremely generous with me—not only did they take the time to explain their work to me, they did so with contagious enthusiasm.

Once again, I wish to thank Wendy Wolf, my editor; Don Homolka, my copyeditor; and my agents, John Brockman and Katinka Matson. I'd also like to thank my friends and loved ones who have shared ideas and who have been so supportive, including Oliver Morton, David Harris, Meridith Walters, and, of course, my brother, mother, and father. Thank you all.

INDEX

Page numbers in *italics* refer to illustrations.